Simulating Potential Structural and Operational Changes for Detroit Dam on the North Santiam River, Oregon, for Downstream Temperature Management

By Norman L. Buccola, Stewart A. Rounds, Annett B. Sullivan, and John C. Risley

Prepared in cooperation with the U.S. Army Corps of Engineers

Scientific Investigations Report 2012–5231
Version 1.1, June 2013

U.S. Department of the Interior
U.S. Geological Survey

U.S. Department of the Interior
KEN SALAZAR, Secretary

U.S. Geological Survey
Marcia K. McNutt, Director

U.S. Geological Survey, Reston, Virginia: 2012
Revised: 2013

For more information on the USGS—the Federal source for science about the Earth, its natural and living resources, natural hazards, and the environment, visit http://www.usgs.gov or call 1–888–ASK–USGS.

For an overview of USGS information products, including maps, imagery, and publications,
visit http://www.usgs.gov/pubprod

To order this and other USGS information products, visit http://store.usgs.gov

Suggested citation:
Buccola, N.L, Rounds, S.A., Sullivan, A.B., and Risley, J.C., 2012, Simulating potential structural and operational changes for Detroit Dam on the North Santiam River, Oregon, for downstream temperature management: U.S. Geological Survey Scientific Investigations Report 2012–5231, 68 p.

Contents

Contents—Continued

Figures

Figures—Continued

Figures—Continued

Tables

Conversion Factors, Datums, and Abbreviations and Acronyms

Conversion Factors

Inch/Pound to SI

Multiply	By	To obtain
foot (ft)	0.3048	meter (m)
mile (mi)	1.609	kilometer (km)
square mile (mi^2)	2.590	square kilometer (km^2)
acre-foot (acre-ft)	1,233	cubic meter (m^3)
foot per second (ft/s)	0.3048	meter per second (m/s)
cubic foot per second (ft^3/s)	0.02832	cubic meter per second (m^3/s)
square foot per second (ft^2/s)	0.0929	square meter per second (m^2/s)

SI to Inch/Pound

Multiply	By	To obtain
meter (m)	3.281	foot (ft)
cubic meter per second (m^3/s)	70.07	acre-foot per day (acre-ft/d)
square meter per second (m^2/s)	10.7639	square foot per second (ft^2/s)

Temperature in degrees Celsius (°C) may be converted to degrees Fahrenheit (°F) as follows:
$$°F=(1.8×°C)+32.$$

Temperature in degrees Fahrenheit (°F) may be converted to degrees Celsius (°C) as follows:
$$°C=(°F-32)/1.8.$$

Datums

Vertical coordinate information is referenced to the North American Vertical Datum of 1988 (NAVD 88).

Horizontal coordinate information is referenced to the North American Datum of 1983 (NAD 83).

Elevation, as used in this report, refers to distance above the vertical datum.

Abbreviations and Acronyms

Abbreviation or acronym	Description
7dADM	7-day moving average of the daily maximum
ATU	Accumulated Thermal Unit
BiOP	Biological Opinion
CE-QUAL-W2	2-dimensional hydrodynamic and water-quality model
DMR	Discharge Monitoring Report
MAE	mean absolute error
ME	mean error
NCDC	National Climatic Data Center (National Oceanic and Atmospheric Administration)
NS	Nash-Sutcliffe coefficient
ODEQ	Oregon Department of Environmental Quality
RAWS	Remote Automated Weather Station
RM	river mile
RMSE	root mean square error
SRML	Solar Radiation Monitoring Laboratory (University of Oregon)
USGS	U.S. Geological Survey

Simulating Potential Structural and Operational Changes for Detroit Dam on the North Santiam River, Oregon, for Downstream Temperature Management

By Norman L. Buccola, Stewart A. Rounds, Annett B. Sullivan, and John C. Risley

Executive Summary

Detroit Dam was constructed in 1953 on the North Santiam River in western Oregon and resulted in the formation of Detroit Lake. With a full-pool storage volume of 455,100 acre-feet and a dam height of 463 feet, Detroit Lake is one of the largest and most important reservoirs in the Willamette River basin in terms of power generation, recreation, and water storage and releases. The U.S. Army Corps of Engineers operates Detroit Dam as part of a system of 13 reservoirs in the Willamette Project to meet multiple goals, which include flood-damage protection, power generation, downstream navigation, recreation, and irrigation.

A distinct cycle in water temperature occurs in Detroit Lake as spring and summer heating through solar radiation creates a warm layer of water near the surface and isolates cold water below. Controlling the temperature of releases from Detroit Dam, therefore, is highly dependent on the location, characteristics, and usage of the dam's outlet structures. Prior to operational changes in 2007, Detroit Dam had a well-documented effect on downstream water temperature that was problematic for endangered salmonid fish species, releasing water that was too cold in midsummer and too warm in autumn. This unnatural seasonal temperature pattern caused problems in the timing of fish migration, spawning, and emergence.

In this study, an existing calibrated 2-dimensional hydrodynamic water-quality model [CE-QUAL-W2] of Detroit Lake was used to determine how changes in dam operation or changes to the structural release points of Detroit Dam might affect downstream water temperatures under a range of historical hydrologic and meteorological conditions. The results from a subset of the Detroit Lake model scenarios then were used as forcing conditions for downstream CE-QUAL-W2 models of Big Cliff Reservoir (the small reregulating reservoir just downstream of Detroit Dam) and the North Santiam and Santiam Rivers.

Many combinations of environmental, operational, and structural options were explored with the model scenarios. Multiple downstream temperature targets were used along with three sets of environmental forcing conditions representing *cool/wet, normal,* and *hot/dry* conditions. Five structural options at Detroit Dam were modeled, including the use of existing outlets, one hypothetical variable-elevation outlet such as a sliding gate, a hypothetical combination of a floating outlet and a fixed-elevation outlet, and a hypothetical combination of a floating outlet and a sliding gate. Finally, 14 sets of operational guidelines for Detroit Dam were explored to gain an understanding of the effects of imposing different downstream minimum streamflows, imposing minimum outflow rules to specific outlets, and managing the level of the lake with different timelines through the year. Selected subsets of these combinations of operational and structural scenarios were run through the downstream models of Big Cliff Reservoir and the North Santiam and Santiam Rivers to explore how hypothetical changes at Detroit Dam might provide improved temperatures for endangered salmonids downstream of the Detroit-Big Cliff Dam complex.

Conclusions that can be drawn from these model scenarios include:

- The water-temperature targets set by the U.S. Army Corps of Engineers for releases from Detroit Dam can be met through a combination of new dam outlets or a delayed drawdown of the lake in autumn.

- Spring and summer dam operations greatly affect the available release temperatures and operational flexibility later in the autumn. Releasing warm water during midsummer tends to keep more cool water available for release in autumn.

- The ability to meet downstream temperature targets during spring depends on the characteristics of the available outlets. Under existing conditions, although warm water sometimes is present at the lake surface in spring and early summer, such water may not be available for release if the lake level is either well below or well above the spillway crest.

- Managing lake releases to meet downstream temperature targets depends on having outlet structures that can access both (warm) lake surface water and (cold) deeper lake water throughout the year. The existing outlets at Detroit Dam do not allow near-surface waters to be released during times when the lake surface level is below the spillway (spring and autumn).

- Using the existing outlets at Detroit Dam, lake level management is important to the water temperature of releases because it controls the availability and depth of water at the spillway. When lake level is lowered below the spillway crest in late summer, the loss of access to warm water at the lake surface can result in abrupt changes to release temperatures.

- Because the power-generation intakes (penstocks) are 166 feet below the full-pool lake level, imposing minimum power production requirements at Detroit Dam limits the amount of warm surface water that can be expelled from the lake in midsummer, thereby postponing and amplifying warm outflows from Detroit Lake into the autumn spawning season.

- Likewise, imposing minimum power production requirements at Detroit Dam in autumn can limit the amount of cool hypolimnetic water that is released from the lake, thereby limiting cool outflows from Detroit Lake during the autumn spawning season.

- Model simulations indicate that a delayed drawdown of Detroit Lake in autumn would result in better control over release temperatures in the immediate downstream vicinity of Big Cliff Dam, but the reduced outflows necessary to retain more water in the lake in late summer are more susceptible to rapid heating downstream.

- Compared to the existing outlets at Detroit Dam, floating or sliding-gate outlet structures can provide greater control over release temperatures because they provide better access to warm water at the lake surface and cooler water at depth.

These conclusions can be grouped into several common themes. First, optimal and flexible management and achievement of downstream temperature goals requires that releases of warm water near the surface of the lake and cold water below the thermocline are both possible with the available dam outlets during spring, summer, and autumn. This constraint can be met to some extent with existing outlets, but only if access to the spillway is extended into autumn by keeping the lake level higher than called for by the current rule curve (the typical target water-surface elevation throughout the year). If new outlets are considered, a variable-elevation outlet such as a sliding gate structure, or a floating outlet in combination with a fixed-elevation outlet at sufficient depth to access cold water, is likely to work well in terms of accessing a range of water temperatures and achieving downstream temperature targets.

Furthermore, model results indicate that it is important to release warm water from near the lake surface during midsummer. If not released downstream, the warm water will build up at the top of the lake as a result of solar energy inputs and the thermocline will deepen, potentially causing warm water to reach the depth of deeper fixed-elevation outlets in autumn, particularly when the lake level is drawn down to make room for flood storage. Delaying the drawdown in autumn can help to keep the thermocline above such outlets and preserve access to cold water.

Although it is important to generate hydropower at Detroit Dam, minimum power-production requirements limit the ability of dam operators to meet downstream temperature targets with existing outlet structures. The location of the power penstocks below the thermocline in spring and most of summer causes the release of more cool water during summer than is optimal. Reducing the power-production constraint allows the temperature target to be met more frequently, but at the cost of less power generation.

Finally, running the Detroit Dam, Big Cliff Dam, and North Santiam and Santiam River models in series allows dam operators to evaluate how different operational strategies or combinations of new dam outlets might affect downstream temperatures for many miles of critical endangered salmonid habitat. Temperatures can change quickly in these downstream reaches as the river exchanges heat with its surroundings, and heating or cooling of 6 degrees Celsius is not unusual in the 40–50 miles downstream of Big Cliff Dam.

The results published in this report supersede preliminary results published in U.S. Geological Survey Open-File Report 2011-1268 (Buccola and Rounds, 2011). Those preliminary results are still valid, but *the results in this report are more current and comprehensive*.

Introduction

Detroit Dam was constructed in 1953 by the U.S. Army Corps of Engineers (USACE) on the North Santiam River in western Oregon and resulted in the formation of Detroit Lake (fig. 1). The North Santiam River drains an area on the western slopes of the Cascade Range, and it is one of several major tributaries to the Willamette River (fig. 2). Detroit Dam is the tallest dam (463 ft) in the Willamette River basin and impounds 455,100 acre-ft of water at full pool, making Detroit Lake one of the largest reservoirs in the basin. The smaller reregulating dam downstream of Detroit Dam, Big Cliff Dam, ensures steady streamflows in the North Santiam River and allows Detroit Dam's power generating facility (and releases) to be turned on and off during the course of a day to meet peak electrical demands. The Big Cliff–Detroit Dam complex typically generates more hydroelectric power than any other USACE facility in the Willamette River basin, and Detroit Lake ranks as one of the most important recreational resources among the 13 reservoirs managed by USACE in the Willamette Project.

Prior to 2007, power generation was a high priority for the Big Cliff–Detroit Dam complex, and releases from Detroit Dam generally were routed through the power penstocks (centerline elevation 427.6 m [1,402.9 ft]) except for times when excess flows were released through the upper regulating outlets (ROs, center-line elevation 408.4 m [1,339.9 ft]) or over the spillway (crest elevation 469.7 m [1,541.0 ft]). During those years, midsummer releases were unseasonably cold because the power penstocks are located 166 ft below Detroit Lake's full-pool level, well below the thermocline at that time of year. Releases from that depth allowed summer solar energy inputs to accumulate in a growing layer of warm water at the lake surface. Drawdown of the lake in September to make room for winter flood storage typically brought that warmer water down to the level of the power penstocks, thus resulting in unseasonably warm releases in late summer and autumn. These somewhat "unnatural" seasonal patterns in the temperature releases can be confusing to anadromous fish, altering the timing of migration, spawning, and egg emergence (Caissie, 2006). The thermal effects of Willamette River basin dams have been quantified in recent modeling studies, and the effects can extend for many miles and many days of travel time downstream (Rounds, 2010).

The North Santiam River and its tributaries (fig. 2) provide habitat for endangered Upper Willamette River Chinook salmon (*Oncorhynchus tshawytscha*) and Upper Willamette River winter steelhead (*O. mykiss*). The Oregon Department of Environmental Quality (ODEQ) has set maximum water-temperature standards for stream reaches in Oregon, including the North Santiam and Santiam Rivers, to protect certain life stages of these sensitive fish. These criteria are based on the 7-day moving average of the daily maximum (7dADM) water temperature. For example, the North Santiam River was designated as core cold-water habitat for June 16–August 31 annually, with the 7dADM water temperature not to exceed 16.0°C, and as salmon and steelhead spawning habitat for September 1–June 15, with a stricter 13.0°C criterion. Farther downstream, the Santiam River was designated as salmon and trout rearing and migration habitat with a maximum 7dADM water temperature of 18.0°C for May 16–October 14, and salmon and steelhead spawning habitat for October 15–May 15 with the 13°C maximum criterion for spawning. (Oregon Department of Environmental Quality, 2009).

To protect and enhance these beneficial uses and habitats, the National Marine Fisheries Service wrote a 2008 Willamette Basin Biological Opinion (BiOP) that, among other things, urges the USACE to assess the feasibility of developing project-specific alternatives for achieving long-term temperature control at the Big Cliff–Detroit Dam complex (National Marine Fisheries Service, 2008). The USACE is in the process of evaluating alternatives for both current and long-term downstream temperature management as well as fish passage at many of the dams in the Willamette Project.

Detroit Dam is an excellent facility for the USACE to test strategies for downstream temperature management because the dam has outlets at several fixed elevations, allowing water to be released from multiple depths and blended to meet a downstream temperature target. In particular, the release of warm water over the spillway in midsummer and cool water from deep in the lake in late summer and early autumn can help mitigate problems associated with water temperatures that otherwise are too cold or too warm for fish. Since 2007, USACE has used the spillway and the ROs in addition to the power penstocks to improve downstream fish habitat during the various life stages of endangered salmonid fish species, while at the same time balancing the need to generate hydropower.

To help evaluate potential dam operation strategies and future structural options, the USACE can rely on predictions from several models of the Detroit Lake and North Santiam River system. The U.S. Geological Survey (USGS) has constructed a model of Detroit Lake to examine water-temperature and suspended-sediment conditions in the lake and downstream (Sullivan and others, 2007). The model was built using CE-QUAL-W2, a two-dimensional, laterally averaged hydrodynamic and water-quality model from USACE (Cole and Wells, 2002) that is widely applied to river and reservoir systems around the world. The USGS Detroit Lake model was calibrated to conditions during calendar years 2002 and 2003 and was tested for high-flow conditions in 2005–06. The model and many results are available online at http://or.water.usgs.gov/santiam/detroit_lake/.

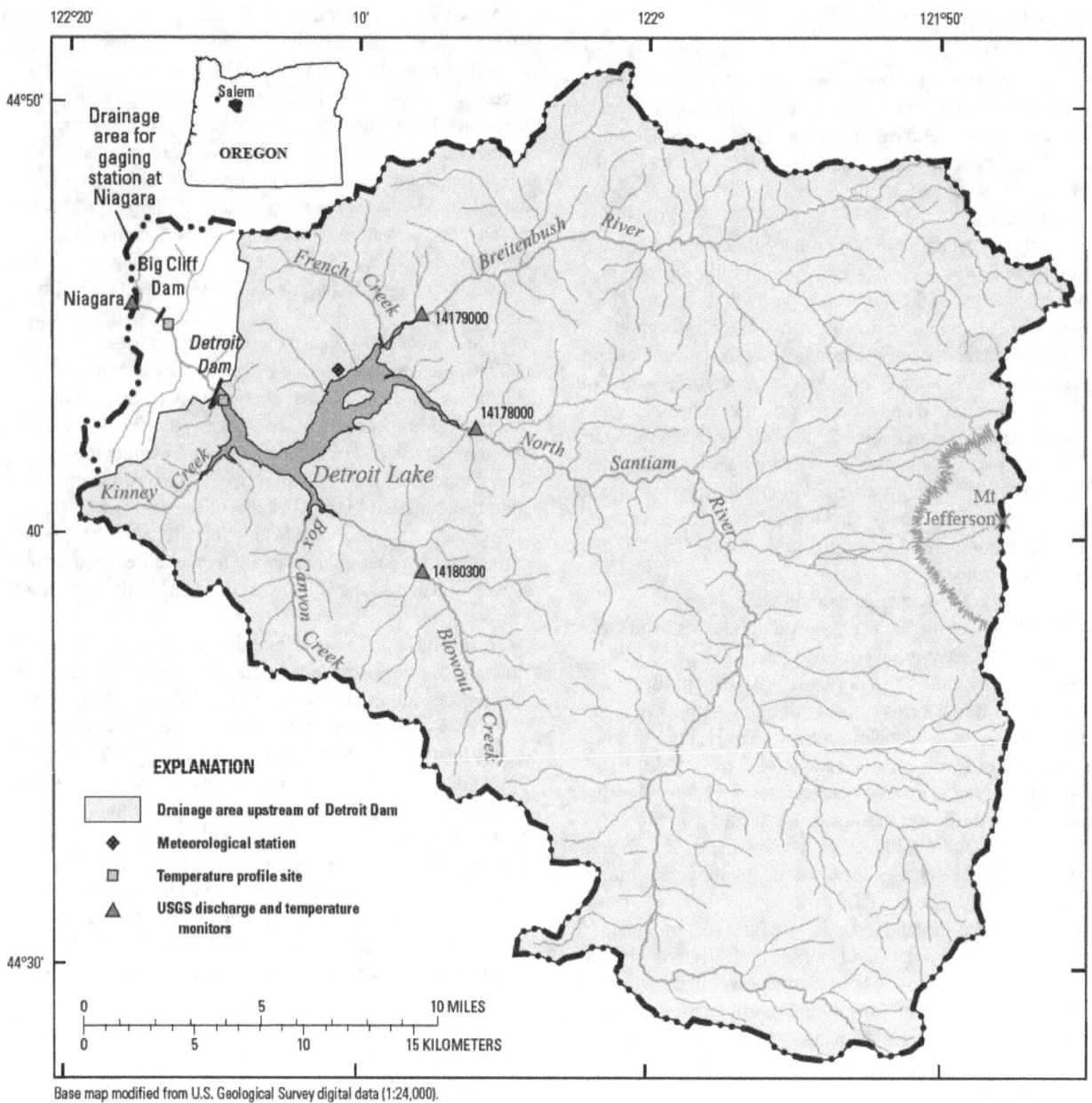

Base map modified from U.S. Geological Survey digital data (1:24,000).
Projection: UTM, Zone 10, North American Datum of 1927.

Figure 1.   Location of Detroit Lake, Detroit Dam, and Big Cliff Dam in the North Santiam River basin in western Oregon. (Map modified from Sullivan and others, 2007).

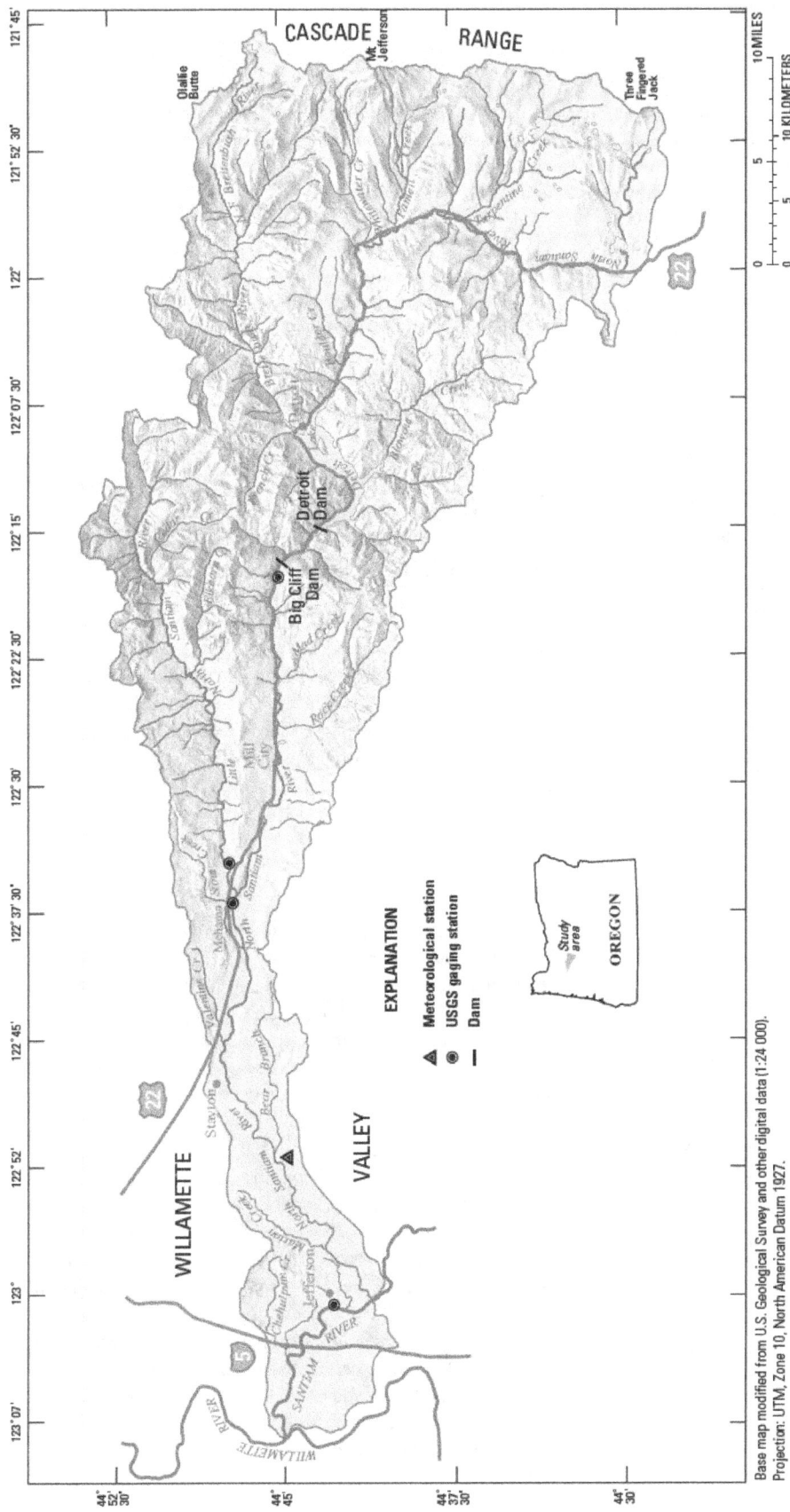

Base map modified from U.S. Geological Survey and other digital data (1:24 000).
Projection: UTM, Zone 10, North American Datum 1927.

Figure 2.   North Santiam and Santiam Rivers and the North Santiam River basin in western Oregon. (Map reproduced from Sullivan and Rounds, 2004).

The USGS Detroit Lake model was built with CE-QUAL-W2 version 3.12, modified to include a custom subroutine that allows a model user to easily estimate release rates from different dam outlets that are necessary to achieve a time series of downstream temperature targets (Rounds and Sullivan, 2006; Buccola and Rounds, 2011). In this way, dam operations can be forecast to meet certain downstream fish habitat criteria at different times of the year. CE-QUAL-W2 models of Big Cliff Reservoir and the North Santiam and Santiam Rivers (Sullivan and Rounds, 2004) also are available. Using those models, predicted flows and water temperatures from the Detroit Lake model can be translated downstream to evaluate how temperatures change in the 61 mi of river downstream of Detroit Dam before the Santiam River joins the Willamette River.

Purpose and Scope

To better inform structural and operational planning decisions related to Detroit Dam outflow temperature management, the USACE asked the USGS to assist in temperature modeling of the Detroit Lake–Big Cliff Reservoir–North Santiam River system. The purpose of this report is to provide water temperature estimates throughout the North Santiam River system from just upstream of Detroit Lake to the junction of the North and South Santiam Rivers (49.2 mi downstream of Detroit Dam) under a range of environmental conditions, alternative dam operations, and potential structural changes at Detroit Dam. Model results presented in this report are intended to inform the current and future operation of Big Cliff and Detroit Dams (and other similar dams in the Pacific Northwest) as well as the planning process for potential structural alterations to Detroit Dam undertaken by USACE for the purpose of improving downstream temperature conditions for fish in the North Santiam River.

The following guiding objectives were used to examine and quantify the downstream thermal effects of potential operational and structural changes to Detroit Dam:

- Develop a range of environmental conditions that represent "*cool/wet*," "*normal*," and "*hot/dry*" hydrologic and meteorological inputs that can serve as boundary conditions for all scenarios.

- Estimate water temperatures in the North Santiam River that might occur in the absence of dams.

- Simulate a range of potential operational and structural scenarios at Detroit Dam and compare predicted outflow temperatures against existing conditions.

- Simulate conditions downstream of Detroit Dam using the Big Cliff Reservoir and North Santiam River models for a select subset of model scenarios and compare to existing conditions.

This study used previously developed CE-QUAL-W2 models of Detroit Lake (Sullivan and others, 2007), Big Cliff Reservoir (model development documented in appendix A), and the North Santiam River (Sullivan and Rounds, 2004) for all simulations of water discharge and temperature. After an assessment of variations in historical data, measured meteorological and hydrologic data from 2002, 2005, 2006, and 2009 were used in this study for forcing conditions to the models and calculations. The calibration performance of the Detroit Lake and Big Cliff Reservoir models was checked using existing operating conditions in the 2011 calendar year, a year in which measured temperature profiles existed in both lakes. By using measured data from 2002 to 2011, the simulations reflect the most current climatic conditions and take advantage of the extensive datasets collected in recent years.

Methods

Flow and Temperature Models

Three separate CE-QUAL-W2 models were used in this study to simulate Detroit Lake, Big Cliff Reservoir, and the North Santiam and Santiam Rivers. All of the model scenarios presented in this report were simulated using the Detroit Lake model, but only some of these scenarios were run with the downstream models of Big Cliff Reservoir and the North Santiam and Santiam Rivers.

Detroit Lake and the Custom Blending Routine

The CE-QUAL-W2 version 3.12 model of Detroit Lake was developed originally by Sullivan and others (2007) for conditions that occurred primarily in 2002 and 2003 when the primary outlet structure was the power penstocks. Since 2007, typical operations at Detroit Dam include releases through the power penstocks as well as seasonal usage of the spillway (during summer) and the upper ROs (during autumn) for downstream temperature management. To ensure that the previously calibrated model accurately represented these new dam operations, the Detroit Lake model calibration was checked and updated using conditions from calendar year 2011 before it (the Detroit Lake model) was applied to the other scenarios of this project. See appendix B for a description of model performance and slight alterations in the calibration for 2011. In general, the model was accurate in its depiction of 2011 in-lake vertical temperature profiles and release temperatures, with mean errors showing a slight negative bias, but within 0.5°C, and mean absolute errors less than about 0.8°C for the profiles and about 0.9°C for the releases, in good agreement with previously documented model performance for Detroit Lake (Sullivan and others, 2007).

The previously developed Detroit Lake model used in this study already included a custom subroutine designed to optimize releases from a set of user-specified outlets to meet a downstream temperature target (Rounds and Sullivan, 2006; Sullivan and Rounds, 2006). The user specifies the total release rate time series for a group of potential outlets along with a time series of desired temperature releases. The subroutine then selects two outlets from among the group of usable outlets, and determines the optimal release rates from those outlets that are required to match the user-specified downstream temperatures. The blending algorithm allows the user to specify several types of outlets, including floating, sliding-gate (variable-elevation), and fixed-elevation outlet structures.

For this study, the custom blending algorithm was further modified and improved in several ways. First, the user can specify that a minimum fraction of the total releases be assigned to a particular outlet. This allows, for example, the user to specify that at least 40 percent of the releases from Detroit Dam go through the powerhouse. That capability was used in several of the scenarios in this study. Second, a minimum release rate also can be assigned to a particular outlet. In this study, this feature was used to set a minimum release rate of 400 ft^3/s from a hypothetical floating outlet. Third, the user can specify a priority ranking for each of the outlets in an outlet group, such that one outlet is preferred for releases when (1) the lake is isothermal and the choice of outlet has little to no effect on release temperatures, or (2) the minimum flow fraction or minimum release rate criteria are in conflict. The priority ranking allows the user to assign more flow to power generation, for example, when the lake is isothermal. Finally, the blending algorithm itself was improved, incorporating an iterative solution method that greatly improved the algorithm's ability to match the user-specified temperature target. Because the release temperature from each outlet is a function of flow, an iterative process is required to find the best combination of flows from two different outlets to match the downstream temperature target.

These updates to the blending subroutine require two new input parameters in the model control file, one for the minimum fraction (0 to 1) or minimum release rate (input as a negative number), and a second for the priority ranking. The code changes and updates are described in appendix C. The code changes were meant to be as general as possible, but in this study it was not necessary to specify more than two available outlets at any time. The algorithm for choosing two outlets from among a group of more than two available outlets was not updated to use the priority ranking or the minimum fraction or flow release criteria.

Big Cliff Reservoir

The customized version 3.12 CE-QUAL-W2 executable used for the Detroit Lake temperature model also was used for the Big Cliff Reservoir model. Big Cliff Reservoir is a small reregulating reservoir just downstream of Detroit Dam, and its operation has a small effect on water temperature at some times of the year. The development of the Big Cliff model is discussed in more detail in appendix A.

North Santiam/Santiam River

The CE-QUAL-W2 model used to simulate streamflow and water temperature in the North Santiam and Santiam Rivers from Big Cliff Dam (river mile [RM] 58.1) to the Willamette River confluence was constructed and calibrated in a previous study to support the Willamette River water-temperature Total Maximum Daily Load (TMDL) process (Sullivan and Rounds, 2004). That study relied on version 3.12 of CE-QUAL-W2 but included some customized outputs to help compute water-temperature statistics for the TMDL work. The same model version and configuration used in the previous study was applied in this study. The model grid consisted of 1 water body, 6 branches, and 310 segments extending over approximately 58.1 mi of the North Santiam and Santiam Rivers (fig. 3). A further description of the methods and assumptions involved in setting up and applying this model is in appendix D.

Some of the model scenarios included in this study resulted in temperature releases from Detroit Dam that were either not very successful in matching the intended downstream target or closely matched those of another scenario. Appendix E provides documentation for that set of model scenarios.

Environmental Scenarios

Three distinctly different environmental forcing scenarios—streamflow input, temperature inflow, and weather conditions—were developed to evaluate temperature management operations and structural options at Detroit Dam. To ensure that the streamflow, water temperature, and meteorological datasets used to drive the models were consistent with one another, the simplest approach was to use historical datasets that represented a wide range of possible conditions, from cold and wet to normal to warm and dry.

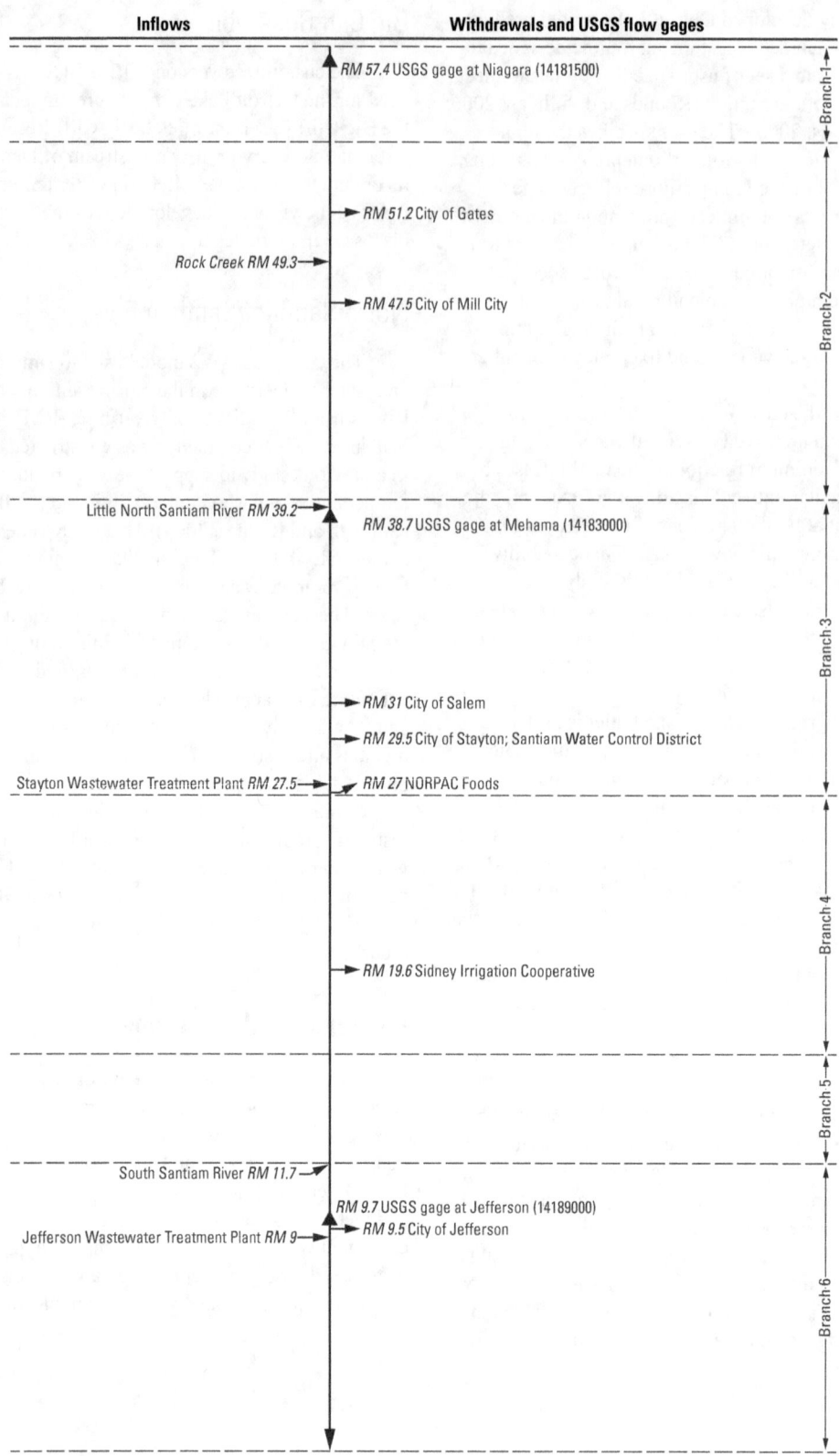

Figure 3.   Locations of point-source and tributary inflows, withdrawals, and USGS streamflow-gaging stations in the North Santiam and Santiam Rivers downstream of Big Cliff Dam, Oregon. (Diagram reproduced from Sullivan and Rounds, 2004.)

The historical data were analyzed based primarily on the assumption that streamflow, along with meteorological conditions, is one of the most important factors influencing stream temperatures in Detroit Lake, Big Cliff Reservoir, and the North Santiam River. In many years, above-average streamflow (driven by snowmelt) during April–June can translate into above-average streamflow during July–September; therefore, the timing of runoff from snowmelt and precipitation may affect midsummer temperatures, and the development of these environmental forcing scenarios must take this relationship into account.

Because streamflow and water temperature typically exhibit less year-to-year variability in late summer (August–September) prior to the autumn rainy season, and because years with a wet winter and spring do not necessarily have a wet autumn, the historical data were divided and analyzed in two periods: "winter–summer" (January–September) and "autumn" (October–December). In this way, measured data from a year with a dry (or wet) winter–summer could be concatenated with measured data from a year with a dry (or wet) autumn, producing a more-or-less uniformly dry (or wet) environmental scenario for modeling that has streamflow, water-temperature, and meteorological data that are consistent with one another. Dividing the year at the beginning of October not only made it easier to splice and transition model input data from separate years, but also takes advantage of the fact that autumn streamflow conditions, once the rainy season begins, are largely independent of the snowpack from the preceding winter and spring.

To select scenarios with the most realistic range in possible streamflow and water temperature throughout the year, a method was devised to rank 10 recent years in which adequate streamflow, water-temperature, and meteorological data were available (2000–2009), using monthly mean flow and temperature data from the North Santiam River below Boulder Creek site upstream of Detroit Lake (USGS site 14178000). In order to avoid a high-flow bias in the monthly flow comparisons, the monthly streamflow was log-transformed prior to computing a difference between each month's flow and the long-term monthly median streamflow. This method allows the low-flow months to be compared more equally with high-flow months, and the differences between years can be assessed more clearly. To rank a group of months in each year, the sum of the differences between the log-transformed monthly mean streamflow and the log-transformed median of the monthly mean streamflow over the entire period of record (1929–2009) was computed and compared for the years 2000–2009. Results for the January–September and October–December time frames are shown in table 1. The same procedure was applied to stream temperature data from the same site using a historical period of record of 1975–2009. This ranking procedure was used to guide further exploration of the hydrologic conditions that occurred in each year.

The rankings in table 1 and a visual comparison of the monthly data were used to develop three scenarios representing *cool/wet*, *normal*, and *hot/dry* conditions based primarily on the rankings for streamflow. For example, the *normal* scenario was created by concatenating data from January 1 to September 27, 2006, with data from September 27 to December 31, 2009 (table 2). Concatenation dates in table 2 were selected based on the day in which streamflow and meteorological conditions closely matched in the 2 years of interest. Streamflow and stream temperature during the three selected scenarios *(cool/wet, normal, and hot/dry)* are shown in figures 4 and 5. Because a large amount of variation in streamflow historically occurs during January to September, the three "winter–summer" scenarios were differentiated primarily by the quantity of streamflow occurring during the spring snowmelt period.

Table 1. Ranking of streamflow and water-temperature conditions at USGS gaging station 14178000 (North Santiam River below Boulder Creek, Oregon) for two periods in each calendar year, 2000–2009.

[Ranks were calculated as log(monthly mean streamflow)–log(median monthly streamflow over entire period of record) and log(monthly mean temperature)–log(median monthly temperature over entire period of record). Low ranks for streamflow indicate lower flows; low ranks for temperature indicate lower temperatures. Colors indicate months that were concatenated for three scenarios: cool/wet = blue; normal = purple; and hot/dry = red]

Year	Spring–Summer		Autumn–Winter	
	Streamflow	Temperature	Streamflow	Temperature
2000	7	4	2	2
2001	1	8	7	9.5
2002	9	3	1	6
2003	5	10	3	7
2004	4	9	4	9.5
2005	2	7	8	5
2006	6	5	10	8
2007	3	6	9	4
2008	10	1	5	3
2009	8	2	6	1

Table 2. Description of environmental scenarios, North Santiam River, Oregon.

Environmental forcings	Measured time-frame		Concatenate date (month-day)
	Spring/Summer	Autumn/Winter	
cool/wet	2009	2006	10-12
normal	2006	2009	09-27
hot/dry	2005	2002	09-27

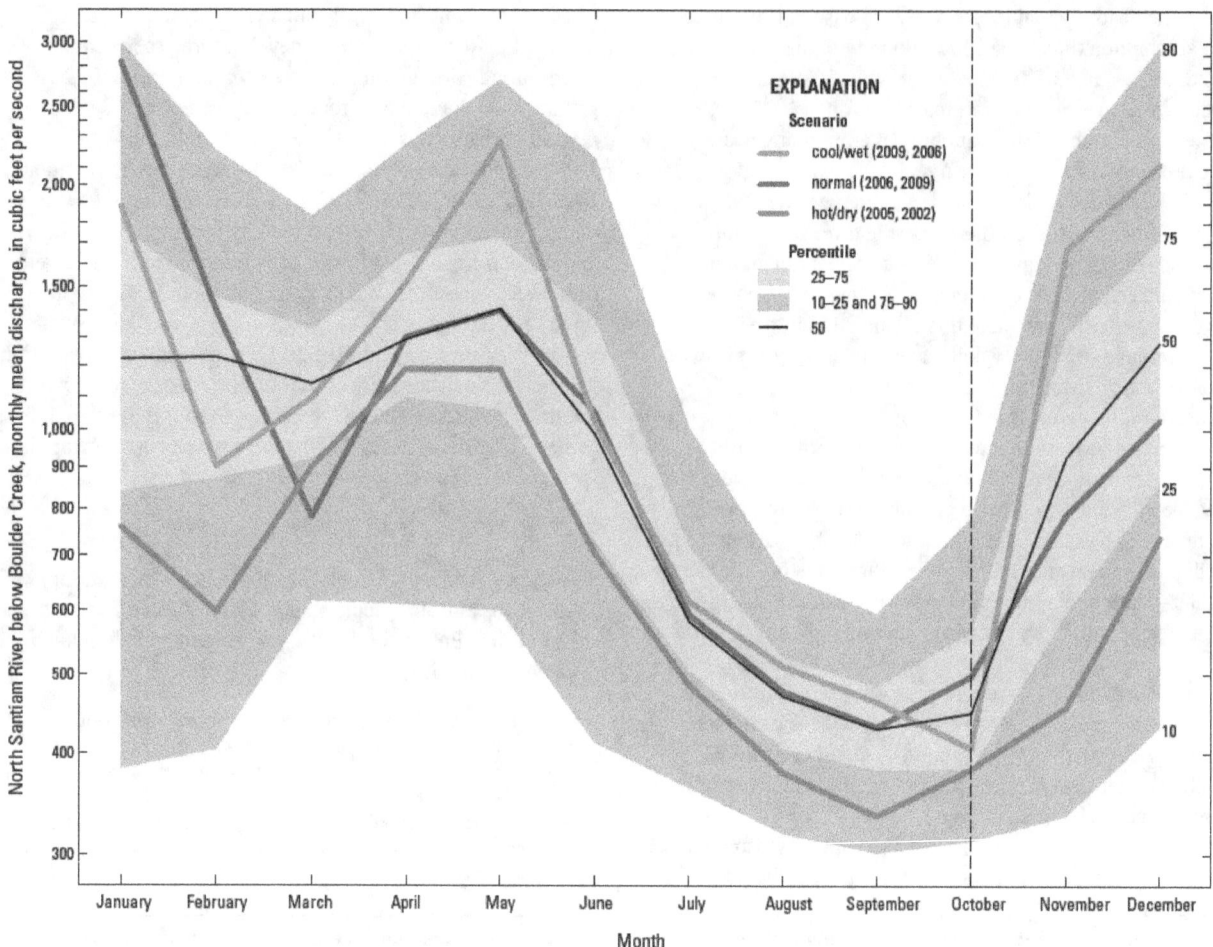

Figure 4. Monthly mean discharge in the North Santiam River below Boulder Creek (USGS station 14178000) under three scenarios, North Santiam River, Oregon. The calendar years in the explanation parentheses denote the 2 years from which data were drawn and concatenated for the January–September and October–December periods.

Together, the three environmental forcing scenarios span more than the 25th–75th percentiles (interquartile range) of the historical data and do not exceed the 10th or 90th percentiles (the central 80 percent of the data used to indicate skewness) of streamflow and temperature. These environmental scenarios, therefore, encompass much of the typical variability in streamflow and water temperature, but without including rare and extreme hydrologic conditions. Most importantly, the *normal* scenario is very near the median streamflow for much of the year aside from January, March, and December.

Streamflow under the *hot/dry* scenario is near the 25th percentile for the entire year, whereas monthly mean stream temperature is above the median for the entire year except for October and November. The result is a warm and dry scenario.

Aside from February and October, monthly mean streamflow under the *cool/wet* scenario is above the median for the entire year. Interestingly, the extremely high flows occurring during autumn of the *cool/wet* scenario correspond to above average stream temperatures (probably due to direct rainfall-runoff), whereas the high flows occurring earlier in the year produced below average stream temperatures (probably due to snowmelt). These results confirm the dependence of North Santiam River stream temperatures on snowmelt from the Cascade Range. Farther downstream, however, river temperatures will depend greatly on dam operations and meteorological conditions.

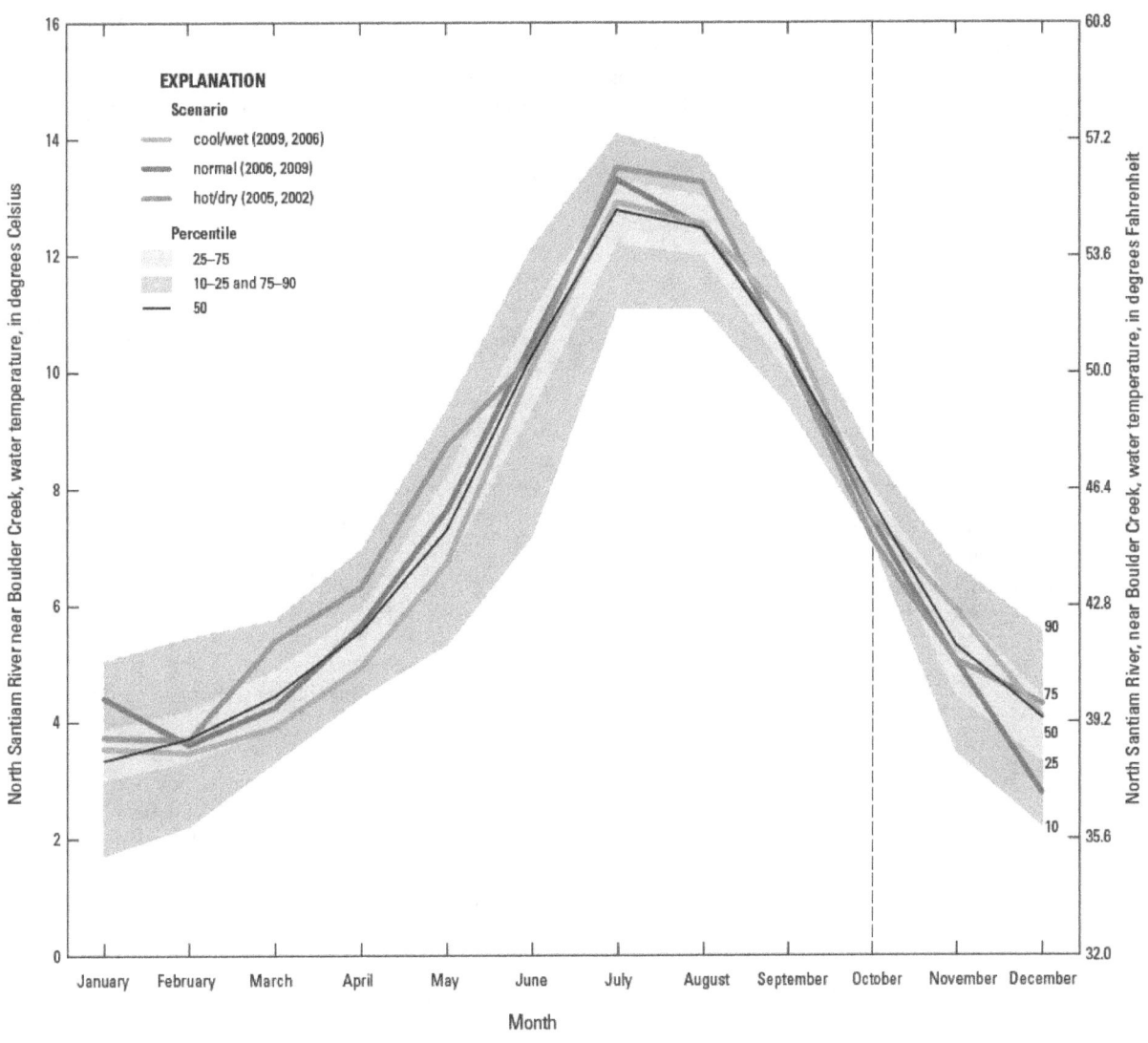

Figure 5. Monthly mean stream temperature in the North Santiam River below Boulder Creek (USGS site 14178000) under three scenarios, North Santiam River, Oregon. The years in the explanation parentheses denote the 2 years from which data were drawn and concatenated for the January–September and October–December periods.

Without-Dam Water Temperature Estimation

Hourly water temperatures for the North Santiam River at Detroit Dam (RM 60.9) were estimated for a "no-dams" scenario, in which Detroit Dam does not exist, for the *hot/dry*, *normal*, and *cool/wet* environmental scenarios. The estimates were computed using a simple mass and energy balance approach combined with a nominal downstream warming rate applied during summer, following methods documented by Rounds (2010). Although a simple one-dimensional model could have been constructed and applied to estimate without-dam water temperatures, a rigorous model was not necessary because these estimates were used only to provide a context for the results of the larger study. With a maximum estimated error of 0.5–0.8°C (Rounds, 2010) over a relatively short 9-mi reach, the simple mass and energy balance method also is likely to be just as accurate and much easier to develop and apply than a one-dimensional model.

The mass and energy balance method was relatively easy to apply because the three major streams entering Detroit Lake have continuous data collection for both streamflow and water temperature during the period 2000–11. Monitoring sites on these three streams are the North Santiam River below Boulder Creek near Detroit (station 14178000), Breitenbush River above French Creek near Detroit (station 14179000), and Blowout Creek near Detroit (station 14180300) (see fig. 1). Applying a mass and energy balance to mix these threes streams together produces the following equation:

$$T_{est} = (Q_{NS}T_{NS} + Q_{BB}T_{BB} + Q_{BL}T_{BL})/(Q_{NS} + Q_{BB} + Q_{BL}) \quad (1)$$

where

T_{est} is mixed water temperature estimate in degrees Celsius,

Q_{NS} is measured streamflow in the North Santiam River at station 14178000 in cubic feet per second,

T_{NS} is measured water temperature in the North Santiam River at station 14178000 in degrees Celsius,

Q_{BB} is measured streamflow in the Breitenbush River at station 14179000 in cubic feet per second,

T_{BB} is measured water temperature in the Breitenbush River at station 14179000 in degrees Celsius,

Q_{BL} is measured streamflow in Blowout Creek at station 14180300 in cubic feet per second, and

T_{BL} is measured water temperature in Blowout Creek at station 14180300 in degrees Celsius.

These hourly water-temperature estimates then were adjusted to account for the instream warming that may occur as water traverses the 9-mi reach between the upstream end of Detroit Lake (where the three tributaries were assumed to join and mix) and Detroit Dam. From November 1 to April 13, or any time of the year when water temperatures were less than 6°C, no adjustments were made to the hourly water temperature estimates. From April 14 to October 31, a downstream warming rate was applied as a function of the mixed temperature estimate, based on an assumption that warmer water was an indication of conditions that were favorable for some warming. All hourly water temperature estimates greater than 14°C were increased by 0.99°C to account for a nominal maximum downstream warming rate of 0.11°C/mi over the 9-mi reach. This maximum downstream warming rate was based on historical data (Moore, 1964, 1967) as well as previous water-temperature modeling in the

North Santiam River in the 4 mi just downstream of Big Cliff Dam (Rounds, 2010). Water-temperature estimates less than 14°C but greater than 6°C were increased to account for some downstream warming, but less than the maximum rate of 0.11°C/mi, using the following linear interpolation:

$$T_{final} = T_{est} + 0.99(T_{est} - 6.0)/(14.0 - 6.0), 6.0 \leq T_{est} \leq 14.0 \quad (2)$$

where T_{final} is the final interpolated water temperature estimate in degrees Celsius.

These without-dam water-temperature estimates for the three environmental scenarios show a more "natural" seasonal temperature pattern that peaks in July or August, in contrast to the pre-2007 downstream temperature peak in September or October (fig. 6). The contrast in the seasonal temperature pattern is more evident in the *hot/dry* and *normal* environmental scenarios because the measured downstream temperatures at the Niagara gage (USGS station 14181500) for those scenarios came primarily from years prior to the operational changes that occurred in 2007 to more actively manage the temperature releases from Detroit Dam. The difference is less evident for the *cool/wet* environmental scenario because reservoir operations through September in that scenario were different and were influenced by a desire to better manage downstream temperatures.

A comparison of mean annual water temperatures with or without the dams showed a negligible difference (less than 0.5°C, table 3). Therefore, the main effect of the dams is to change the seasonal pattern in downstream temperatures and the timing and magnitude of the annual maximum and minimum, rather than to increase or decrease temperatures overall.

Temperature Targets

Temperature targets for reservoir releases were previously developed and used by USACE for the McKenzie River system downstream of another dam (Cougar Dam on the South Fork McKenzie River) to support a restoration of uses by endangered fish (U.S. Army Corps of Engineers, 2012). Because no fish-based reservoir release targets had been developed for the North Santiam River, and because the endangered fish species are similar in the two systems, the McKenzie River temperature targets were applied in this study. ODEQ had developed some temperature targets for reservoir releases as part of the Willamette River water-temperature TMDL (Oregon Department of Environmental Quality, 2006), but those targets were based on an analysis of the historical data rather than the needs of the fish. Similarly, the maximum temperature criteria embedded in Oregon's temperature standard, while meant to be protective for fish, include less seasonal variation than the USACE targets.

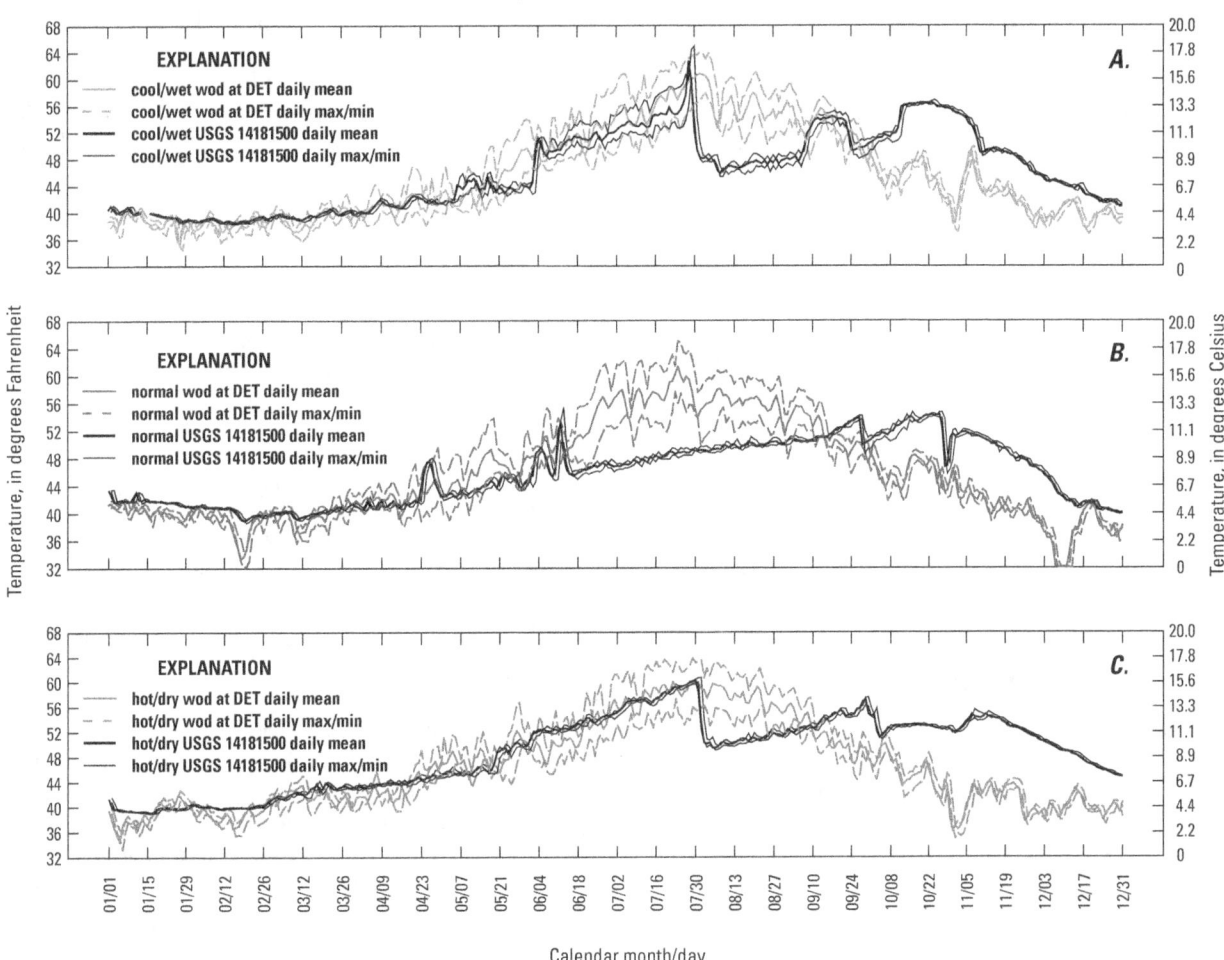

Figure 6. Estimated without-dam water temperatures (labeled "wod") at the Detroit Dam site (DET) compared to measured (USGS station 14181500) hourly water temperatures for the three environmental scenarios, North Santiam River, Oregon. (*A*) *cool/wet* (2009 spliced with 2006), (*B*) *normal* (2006 spliced with 2009), and (*C*) *hot/dry* (2005 spliced with 2002).

Table 3. Mean, minimum, and maximum annual without-dam temperature estimates (labeled "W/o-dams") at Detroit Dam (river mile 60.9) along with measured mean, minimum, and maximum annual with-dam temperatures (labeled "With-dams") at USGS gaging station 14181500 (river mile 57.3) for three environmental scenarios, North Santiam River, Oregon.

[*Measured 14181500 river mile 57.3: Missing period January 12–16, 2009]

	Water temperature, in degrees Celsius					
	Cool/wet		Normal		Hot/dry	
	W/o-dams	With-dams*	W/o-dams	With-dams	W/o-dams	With-dams
Mean	7.46	7.92	7.60	7.64	7.96	8.15
Minimum	1.37	3.50	0.03	3.70	0.71	4.50
Maximum	17.83	18.30	18.36	13.00	17.69	13.70

Therefore, using the USACE targets seemed to be a reasonable choice. Although the USACE temperature targets include both a minimum and maximum monthly value for much of the calendar year, only the maximum values were used for the majority of model runs in this study (scenario names include "*max*" to denote this practice).

Estimated without-dam water temperatures at the Detroit Dam site also were used as temperature targets for selected model scenarios. Using the without-dam temperatures as release targets provided a way to measure the ability of changes in dam operations or alterations in dam outlets to match temperatures that might exist in the absence of Detroit Dam. The estimated without-dam water temperatures were applied as release targets in model scenarios only after first computing the 7dADM—the statistic used in Oregon's temperature standard. That computation removed much of the daily variation in the without-dam temperature estimates, thus making them more useful as release targets. Model scenarios using these targets are named with "*w/o_dams7dADM*" to denote this target.

Model Setup and Usage

Dam Outflow Estimation

Detroit Dam

Prior to running the Detroit Lake model to simulate operational and structural scenarios at Detroit Dam, the previously developed USGS Detroit Lake model was set up and its calibration checked using measured inflows, outflows, and weather conditions for the entire calendar year of 2011 (see appendix B) and from January 1 to August 30 in each environmental scenario. The only adjustments to the calibrated model parameters from the original model were (1) an additional spillway outlet was added as a "LINE" type structure with a model "WIDTH" of 25 m (82 ft) and an elevation (STR ELEV) of 469.7 m (1,541 ft), and (2) a minor change was made to the wind-sheltering coefficients to better reflect the distribution of wind speed across Detroit Lake. Wind sheltering was decreased from 1.0 to 0.9 and increased from 1.0 to 1.2 for model segments upstream and downstream of the Blowout Creek arm of the lake (model segment 21), respectively.

After the model was set up for each environmental scenario, the difference between measured and modeled forebay elevations in the lake was used to determine the quantity of ungaged inflows and outflows for the lake. An additional model input known as the distributed tributary was created to account for any unmeasured overland flows, evaporation, or groundwater flux not accounted for by other boundary conditions, serving to balance the water budget for the lake. This method worked for the period from January 1 to the concatenation date of the environmental scenario because measured inputs and a corresponding measured lake level were available. From the concatenation date to December 31, however, the environmental scenario switched to inflows from a different year, making measured lake level comparisons impossible. Therefore, a proportion of the inflow from each tributary was used to estimate the magnitude of the distributed tributary from the concatenation date to December 31 of each environmental scenario.

Following the water balance calibration, certain scenarios required that the total release rates (outflows) from Detroit Dam met the following conditions:

1. Releases from Detroit Dam should meet minimum and maximum flow requirements as specified by the BiOP (National Marine Fisheries Service, 2008) (table 4).

2. Computed water levels in Detroit Lake should not exceed the reservoir rule curve for more than 5 days.

3. Use of the power penstock outlets for "power peaking" was assumed to occur during the hours of 0500–1200 and 1400–2200 each day

In reality, "power peaking" pertains only to the power penstock outlets; however, to enable the blending subroutine in CE-QUAL-W2 to determine the optimum balance of releases to meet downstream temperature targets, all outlets had to be placed on this flow schedule and used concurrently. Such concurrent releases might not reflect actual future operations, and the timing of releases (concurrent versus staggered) can have an effect on daily temperature variations immediately downstream; therefore, more detailed modeling may be required to optimize actual dam operations once a reasonable scenario is selected. The power peaking schedule was used only on days in which the daily average release rate was less than 2,472 ft³/s (70 m³/s). This rule helped ease the water balance of the downstream Big Cliff Reservoir model and came closer to the way in which Detroit Dam is operated during large storm events.

For the "*existing*" structural scenario group (use of existing outlets), the computed total release rate was distributed among the available outlets. During times in which the forebay elevation in Detroit Lake was computed to be above the spillway, the total outflow was routed to the spillway and power penstocks, a combination that allows access to warm water near the lake surface (spillway) and cooler water at depth (power penstocks), thus achieving a blend of releases that is best positioned to meet the specified temperature target. When the elevation in the lake was computed to fall below the spillway crest, the only available outlets at Detroit Dam were the power penstocks and the upper ROs. The lower ROs are located below the power penstocks and upper ROs, but usage of those outlets may only be possible at extremely low lake levels and was not assessed in this study. Under the "*base*" operational scenario group, the rules for dam releases that are currently in use by USACE were applied to each environmental scenario.

Table 4. Minimum and maximum Detroit Dam outflow requirements for operational scenarios, North Santiam River, Oregon.

[Details for operational scenarios shown in table 5. Flows are daily mean streamflow, in cubic feet per second (ft^3/s). Altered flows are indicated by the numbers in *italics*. **Other rules incorporated in outflow estimation:** Daily maximum flow, 15,000 ft^3/s; Maximum flow through power penstocks, 2,472 ft^3/s. –, no maximum]

Month/ day	Operational scenario group minimum flows (ft^3/s)					
	base	fixed_ elevation	late_refill	early_dd	delay_dd1	delay_dd2
			Minimum flow			
Jan. 1	1,200	1,200	1,200	1,200	1,200	1,200
Feb. 1	1,000	1,000	1,000	1,000	1,000	1,000
Mar. 1	1,000	1,000	1,000	1,000	1,000	1,000
Apr. 16	1,500	1,500	1,500	1,500	1,500	1,500
May 1	1,580	1,580	1,580	1,580	*880*	1,580
May 16	1,580	1,580	1,580	1,580	*880*	1,580
June 1	1,280	*880*	*580*	*1,280*	*880*	1,280
July 1	1,280	*880*	*580*	*1,280*	*880*	1,280
July 16	1,080	*580*	*580*	*1,080*	*880*	1,080
Sept. 1	1,500	*580*	*580*	*1,500*	*880*	*580*
Oct. 16	1,200	*580*	1,200	1,200	1,200	1,200
Dec. 1	1,200	*580*	1,200	1,200	1,200	1,200
Dec. 31	1,200	*580*	1,200	1,200	1,200	1,200
			Maximum flow			
Jan. 1	–	*15,000*	*15,000*	*15,000*	*15,000*	*15,000*
Sept. 1	3,000	3,000	3,000	3,000	3,000	3,000
Sept. 30	3,000	3,000	3,000	3,000	3,000	3,000
Dec. 31	–	*15,000*	*15,000*	*15,000*	*15,000*	*15,000*

Big Cliff Dam

To simulate outflows at Big Cliff Dam that closely approximated reality and balanced inflows with outflows for Big Cliff Reservoir under a range of conditions, a method of estimating outflows at Big Cliff Dam was developed. The outflow at Big Cliff Dam was assumed to be a moving daily average of the outflow rate from Detroit Dam. The upper and lower bounds of the pool elevation in Big Cliff Reservoir (1,182 and 1,194 ft) made it necessary to add a substantial distributed tributary inflow to the Big Cliff model that accounted for several tributary inflows and balanced inflows with outflows while adhering to these narrow elevation bounds. Initially, a distributed tributary flow rate was calculated by the hourly difference between the calculated outflow from Big Cliff Dam (as described in the previous sentence) and the total inflows to Big Cliff Reservoir. An iterative process then was used to adjust this distributed tributary based on the difference between subsequent modeled water-level elevations and a mean pool elevation of 1,188 ft. This resulted in simulations of Big Cliff Reservoir that both resembled current operating elevation rules and led to simulations with a relatively constant pool elevation (further discussion in appendix B).

The simulated temperatures from the Big Cliff Reservoir model under these operating rules was assessed by a comparison to measured vertical profiles of water temperature from a thermistor string in Big Cliff Reservoir and a comparison of simulated outflow temperatures to measured temperatures at the streamgage downstream of Big Cliff Dam at Niagara (USGS gaging station 14181500) during 2011 (further discussion in appendix B).

Meteorological Inputs

Hourly meteorological input data required for CE-QUAL-W2 include air temperature, dew-point temperature, wind speed, wind direction, short-wave solar radiation, and cloud cover. The same meteorological data in a particular environmental scenario were used as input to all three models of Detroit Lake, Big Cliff Reservoir, and the North Santiam/Santiam River. Air temperature, relative humidity, wind speed, and wind direction data were measured hourly at a Remote Automated Weather Station (RAWS) site near Stayton, Oregon (44° 45' N, 122° 52' W, 155-m [509 ft] elevation). Hourly short-wave solar radiation data were obtained from the University of Oregon's Solar

Radiation Monitoring Laboratory (SRML) at their Eugene monitoring station (44° 2' 60" N, 123° 4' 12" W, 150 m [492 ft] elevation). Daytime cloud-cover data were estimated by comparing computed theoretical solar radiation rates with measured solar radiation rates as described by Sullivan and Rounds (2004). Nighttime cloud-cover data were interpolated from cloud-cover estimates at sunset on one day to corresponding estimates at sunrise on the following day.

Detroit Dam Scenarios and Naming Convention

A range of model scenarios at Detroit Dam were explored to evaluate the potential downstream temperature impacts of altered dam operations as well as hypothetical structural changes at Detroit Dam. Hypothetical dam operation scenarios were developed to evaluate the effects of (1) altering the lake level in autumn (early or delayed drawdown to make room for flood storage), which often necessitated some change to the minimum recommended release rates, and (2) placing specific minimum flow constraints on selected outlets to achieve certain outcomes such as a minimum amount of power generation (table 5). In order to change the seasonal pattern or schedule of lake-level elevations, summertime minimum releases from Detroit Dam had to be decreased to varying degrees (table 4). The set of structural scenarios included the existing outlets as well as the use of new floating (at a fixed depth) or sliding-gate (variable-elevation) outlets either alone, together, or in combination with an existing fixed-elevation outlet (table 6).

Selected operational scenarios were combined with selected structural scenarios, projected onto the three environmental forcing conditions (*cool/wet, normal,* and *hot/dry*) and given a set of temperature target requirements to produce the model scenarios of interest (table 7). The combination of these four conditions—operational scenario, structural scenario, environmental scenario, and temperature target—fully describes the major differences between the model scenarios and provides a consistent naming convention.

Table 5. Operational scenario group descriptions, Detroit Dam, North Santiam River, Oregon.

[ft³/s, cubic foot per second; NA, not applicable]

Operational scenario groups	Minimum outflow rules			Important dates	
	Rules governing total outflow	Minimum outflow to power (percent)	Outflow rule for floating outlet	Refill begins	Drawdown begins
"base"	Existing operational rules following BIOP minimum flow requirements	40	NA	Feb. 1	Sept. 1
"10ppmin"		10	No minimum	Feb. 1	Sept. 1
"20ppmin"		20	No minimum	Feb. 1	Sept. 1
"noppmin"		No minimum	No minimum	Feb. 1	Sept. 1
"nomins"		No minimum	No minimum	Feb. 1	Sept. 1
"400fmin"		No minimum	400 ft³/s minimum	Feb. 1	Sept. 1
"400f"		40	400 ft³/s fixed	Feb. 1	Sept. 1
"fixed_elevation"	Decreased minimum flow requirements June 1–Dec. 31; constant pool elevation of 1,440 feet year-round	40	NA	NA	NA
"late_refill"	Decreased minimum flow requirements during the summer	40	NA	June 1	Sept. 1
"early_dd"	Decreased minimum flow requirements June 1–Sept. 1	40	NA	Feb. 1	Aug. 15
"delay_dd1"	Decreased minimum flow requirements May 1–Oct. 15	40	NA	Feb. 1	Sept. 1
"delay_dd1_noppmin"		No minimum	NA	Feb. 1	Sept. 1
"delay_dd2"	Decreased minimum flow requirements Sept. 1–Oct. 15	40	NA	Feb. 1	Oct. 15
"delay_dd2_noppmin"		No minimum	NA	Feb. 1	Oct. 15

Table 6. Structural scenarios group descriptions, Detroit Dam, North Santiam River, Oregon.

Structural scenario groups	Description of Detroit Dam model outlets	Outlet priority
"existing"	Existing outlets (spillway, power penstocks, and upper regulating outlet gates)	Power
"pp-float"	1 floating outlet + existing power penstocks (1,420-foot elevation)	Power
"uro-float"	1 floating outlet + existing upper regulating outlet gates (1,340-foot elevation)	Regulating outlet
"slider1340"	1 sliding outlet from 1,340-foot elevation to the surface	Sliding
"slider1340-float"	1 floating outlet + 1 sliding outlet from 1,340-foot to the surface	Sliding

Table 7. Specification and naming convention of model scenarios, Detroit Dam, North Santiam River, Oregon.

[**Scenario identifier:** c, cool/wet; n, normal; h, hot/dry; wod, without dam. **Bold** scenarios are located in appendix. Orange scenarios were run in Big Cliff and North Santiam River models]

Temperature target	Structural scenarios	Operational scenarios	Scenario identifier Environmental forcings		
			cool/wet	normal	hot/dry
"w/o_dams7dADM"	*"existing"*	*"base"*	cwod1	nwod1	hwod1
		"noppmin"	cwod2	nwod2	hwod2
"max"	*"existing"*	*"base"*	c1	n1	h1
		"noppmin"	c2	n2	h2
		"fixed_elevation"	c3	n3	h3
		"late_refill"	**c4**	**n4**	**h4**
		"early_dd"	**c5**	**n5**	**h5**
		"delay_dd1"	c6	n6	h6
		"delay_dd1_noppmin"	c7	n7	h7
		"delay_dd2"	c8	n8	h8
		"delay_dd2_noppmin"	c9	n9	h9
	"pp-float"	*"nomins"*	c10	n10	h10
		"10ppmin"	**c11**	**n11**	**h11**
		"20ppmin"	c12	n12	h12
		"400fmin"	c13	n13	h13
	"uro-float"	*"400fmin"*	c14	n14	h14
		"20ppmin"	c15	n15	h15
		"40ppmin"	c16	n16	h16
	"slider1340"	*"base"*	c17	n17	h17
	"slider1340-float"	***"delay_dd2"***	**c18**	**n18**	**h18**
		"400f"	c19	n19	h19

Detroit Dam Operational Scenarios

Detroit Dam Reference Conditions

To compare the operational and structural model scenarios, specific reference conditions (entitled "*base*") were used to represent current operational guidelines and structures in place at Detroit Dam. These conditions then were applied with the model using a set of temperature targets to show the extent to which current operations and structures at Detroit Dam were able to meet those temperature targets under the three environmental scenarios.

base Operational Scenario

This operational scenario was intended to provide simulations that reflect the guidelines, timelines, rules, and understandings currently in place for the operation of Detroit Dam under the environmental scenarios developed for this study (table 2). The current operational rules for Detroit Dam were developed for the existing usable outlet structures (spillways, power penstocks, and upper ROs). As mentioned in the section "Detroit Dam Scenarios and Naming Convention", these rules are based on a combination of current minimum outflows (mandated by the BiOP), maximum outflows, downstream irrigation withdrawals, minimum power production requirements, and a schedule of power-peaking operations (*base* scenario in table 4 and table 5). The majority of operational scenarios required a minimum of 40 percent of the total release rate to be routed through the power penstocks to allow a minimum amount of power generation. This is consistent with a current agreement between USACE and the Bonneville Power Administration, which distributes and markets hydropower from Detroit Dam and many other facilities across the Pacific Northwest. All operational scenarios discussed in this report assigned a higher priority to the power penstocks whenever possible.

Before comparing modeled outflow temperatures, it is helpful to compare the modeled forebay elevations in each of the operational scenarios, as the timing of the rule curve can contribute greatly to the resulting temperature regime in the lake. The *base* operational scenarios generally led to modeled lake levels that closely matched the USACE rule curve during spring and early summer. As the summer progressed into the low-flow months, however, minimum flow requirements typically led to outflows exceeding inflows and a gradual decrease in lake level during mid-July through mid-October (fig. 7).

Modeled temperatures from the *base* operations and *existing* structural scenarios serve as a basis to compare other structural and operational scenario outcomes. In this report,

figures of the *max* temperature target scenario results show both the minimum and maximum temperature targets currently used by USACE for the North Santiam River (and adapted from targets used for the McKenzie River, Oregon), but only the maximum temperature target was used to drive the blending algorithm within CE-QUAL-W2 for these scenarios. In many of the following figures, "percent spill" is defined as the percentage of total flow that was directed to outlets other than the power penstocks; thus, spill includes the total releases from the ROs and the spillway. Outflow temperatures from *base* operational scenarios (fig. 8) did not meet the *max* temperature target during summer months (June–August), whereas omitting the minimum power generation constraint in the *noppmin* scenarios (fig. 9) generally allowed the model to be more successful at meeting the *max* temperature target. By allowing the "percent spill" to exceed 60 percent during the summer months (scenarios *c2*, *n2*, and *h2*), the model was allowed to release more of the warmer outflows from the spillway during summer, which retains some of the cooler, deeper water for release in autumn from the upper RO outlets (fig. 9).

Without-Dams Temperature Target

To assess the potential for *base* operations and *existing* structures at Detroit Dam to match temperatures that might exist in the absence of Detroit Dam, calculated temperatures from the without-dams analysis were smoothed using a 7-day moving average of the daily maximum, then used as a temperature target in the *wo_dams7dADM* scenarios (table 7). The primary difference between the two sets of temperature targets used in this study (*max*, *wo_dams7dADM*) is evident in the allowable summer temperatures. Summer dam operations and release temperatures then help to determine the availability of cool water at the elevation of the available outlets later in autumn and the resulting autumn release temperatures (figs. 8 and 10).

As was noted for the *base* and *noppmin* scenarios with *max* temperature targets (*c1*, *n1*, *h1*, *c2*, *n2*, and *h2*), the removal of minimum flow requirements to the power penstocks resulted in more warm surface water released in midsummer and cooler outflows from the upper ROs in autumn (compare figs. 8 and 9). The same is true when the *wo_dams7dADM* temperature targets were applied (compare figs. 10 and 11). The *max* temperature targets in June, early July, and late August are slightly higher than the *wo_dams7dADM* targets. As a result, midsummer release temperatures from the *cwod2*, *nwod2*, and *hwod2* scenarios were slightly lower than temperatures from the *c2*, *n2*, and *h2* scenarios during the same time frame, thus resulting in comparatively warmer outflow temperatures in autumn.

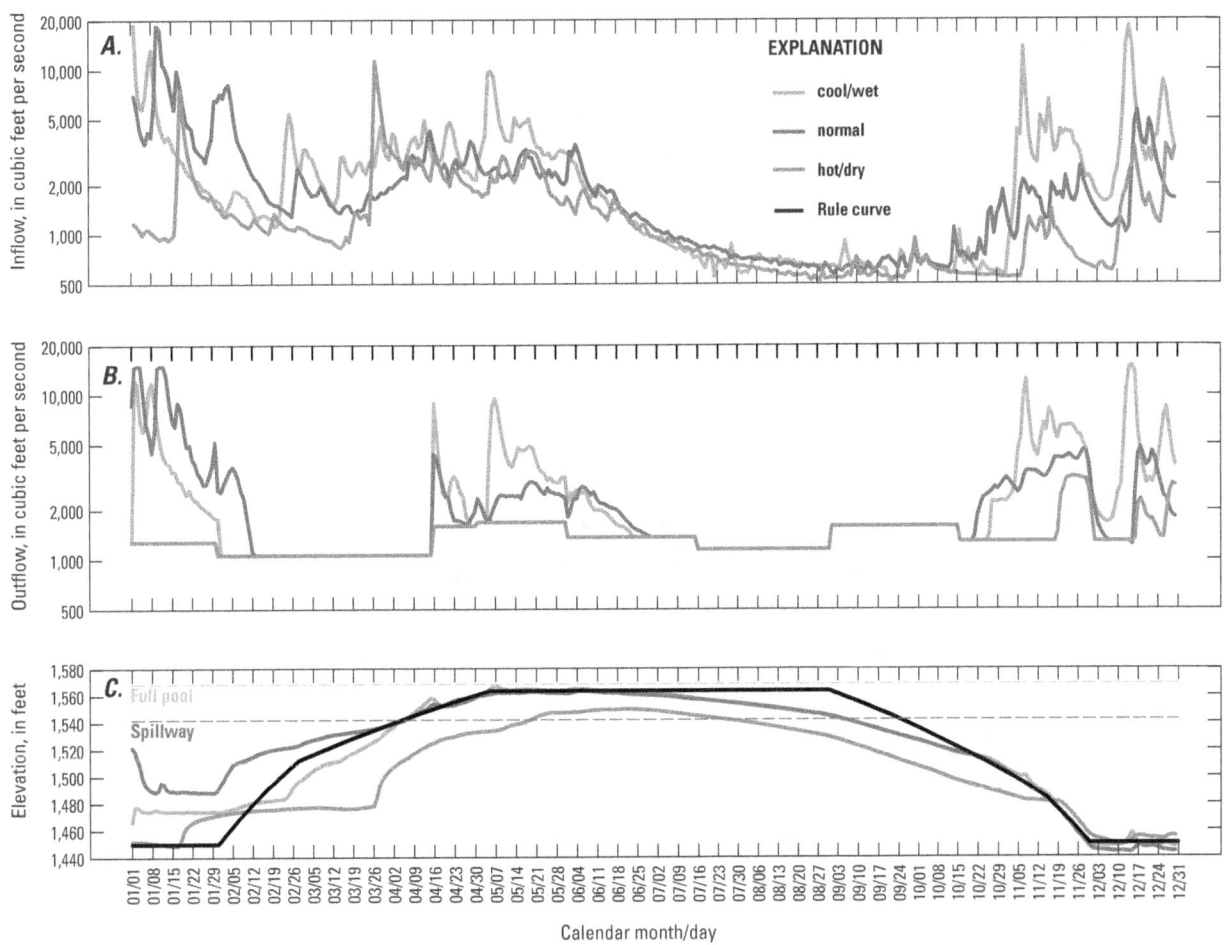

Figure 7. Comparison of *existing* structural scenarios with *base* operational scenarios (scenarios *c1, n1, h1*) (*A*) total inflows, (*B*) total outflows, and (*C*) modeled water-surface elevation and rule curve, North Santiam River, Oregon.

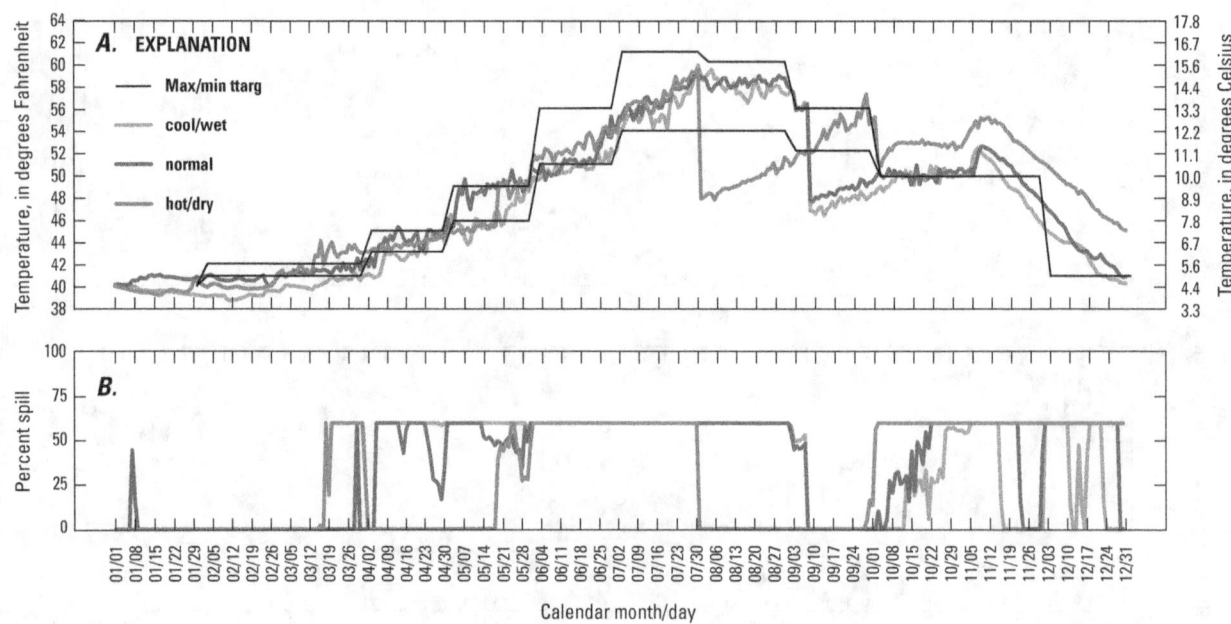

Figure 8. (*A*) Modeled water temperature and (*B*) percent spill for *existing* structural scenarios with *base* operational scenarios, and *max* temperature targets (scenarios *c1*, *n1*, *h1*), North Santiam River, Oregon. Percent spill is the percentage of total flow directed to outlets other than the power penstocks. The maximum and minimum temperature target established for the McKenzie River (labeled "Max/min ttarg") is shown but only the maximum was used in this simulation.

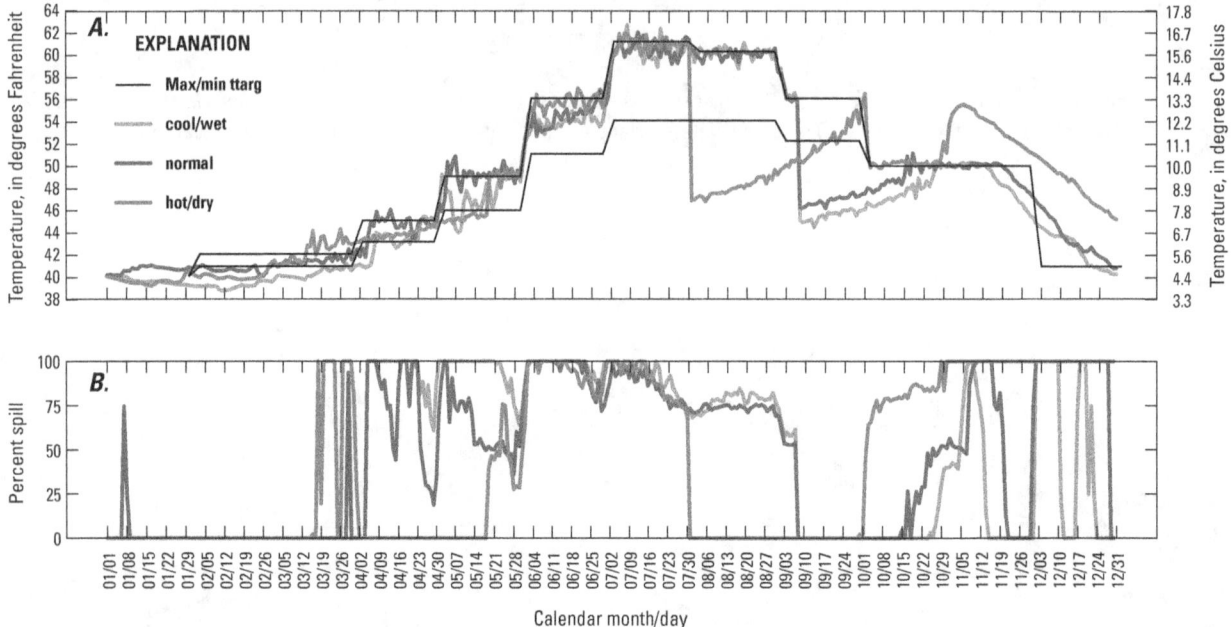

Figure 9. (*A*) Modeled water temperature, and (*B*) percent spill for *existing* structural scenarios with *noppmin* operational scenarios, and *max* temperature targets (scenarios *c2*, *n2*, *h2*), North Santiam River, Oregon. Percent spill is the percentage of total flow directed to outlets other than the power penstocks. The maximum and minimum temperature target established for the McKenzie River (labeled "Max/min ttarg") is shown but only the maximum was used in this simulation.

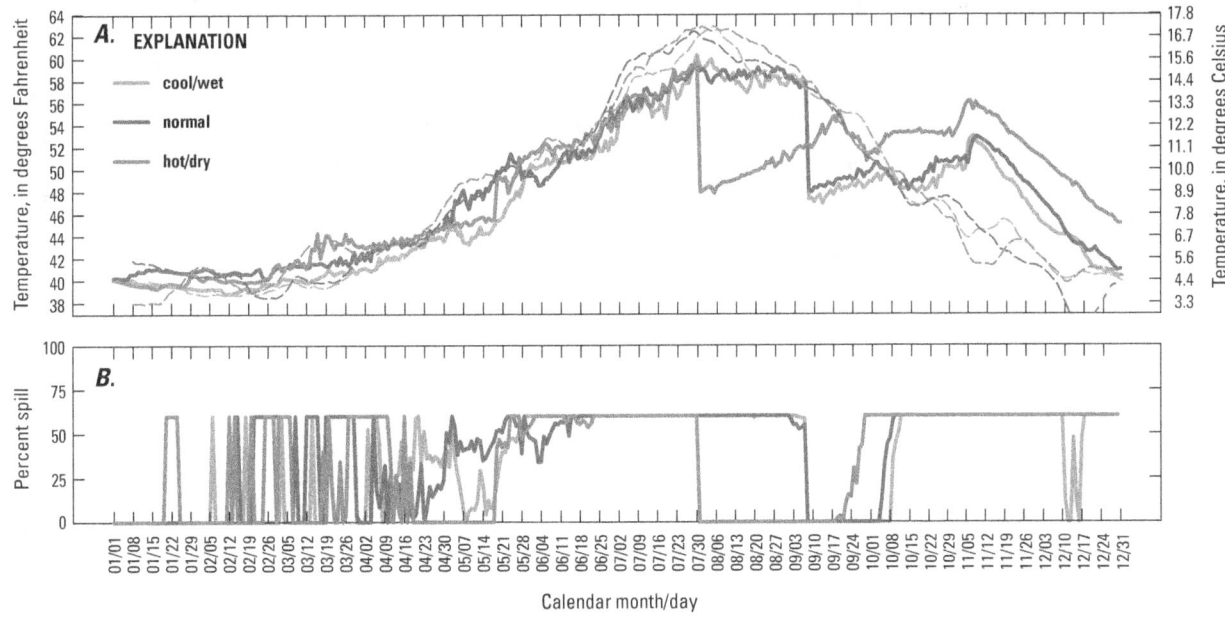

Figure 10. (*A*) Modeled water temperature and (*B*) percent spill for *existing* structural scenarios with *base* operational scenarios and *wo_dams7dADMax* temperature targets (scenarios *cwod1, nwod1, hwod1*), North Santiam River, Oregon. Percent spill is the percentage of total flow directed to outlets other than the power penstocks. The dashed lines correspond to *wo_dams7dADMax* temperature targets in *cool/wet* (blue), *normal* (purple), and *hot/dry* (red) scenarios.

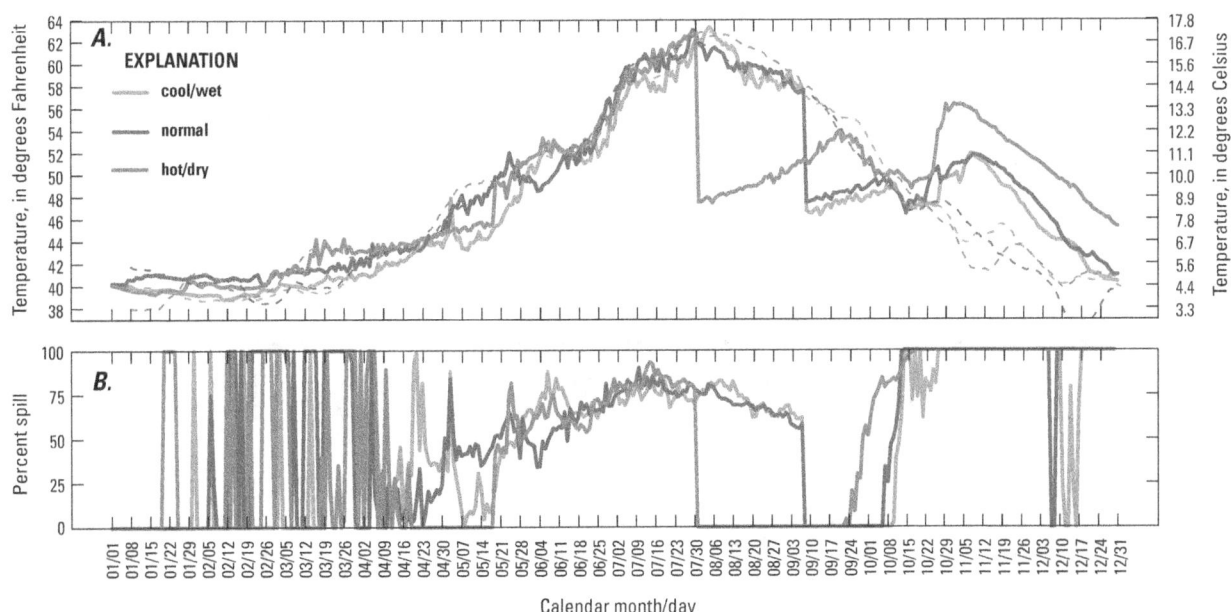

Figure 11. (*A*) Modeled water temperature and (*B*) percent spill for *existing* structural scenarios with *noppmin* operational scenarios and *wo_dams7dADMax* temperature targets (scenarios *cwod2, nwod2, hwod2*), North Santiam River, Oregon. Percent spill is the percentage of total flow directed to outlets other than the power penstocks. The dashed lines correspond to *wo_dams7dADMax* temperature targets in *cool/wet* (blue), *normal* (purple), and *hot/dry* (red) scenarios.

Fixed Lake Level at Minimum Conservation Pool (1,450 feet)

One hypothetical operational scenario included a specification that the year-round Detroit Lake water-surface elevation be held at "minimum conservation pool" the current guideline used during the winter months (fig. 12). This *fixed_elevation* scenario is extreme in that it would greatly affect summer recreational activities on the lake, but might possibly have some advantages for passing fish downstream; this scenario might never be pursued, but resource managers felt it was important to assess the implications of such an action. This scenario also is referred to as "run-of-river" because outflows are approximately equivalent to total inflows throughout the year. Exceptions to this general rule are that outflows cannot fall below the minima or exceed the maxima shown in table 4. Outflow temperatures under *fixed_elevation* scenarios exceeded the *max* temperature target during the months of September–November (fig. 13), mainly because any cool water deep in the lake was below the elevation of the power penstocks and the upper ROs.

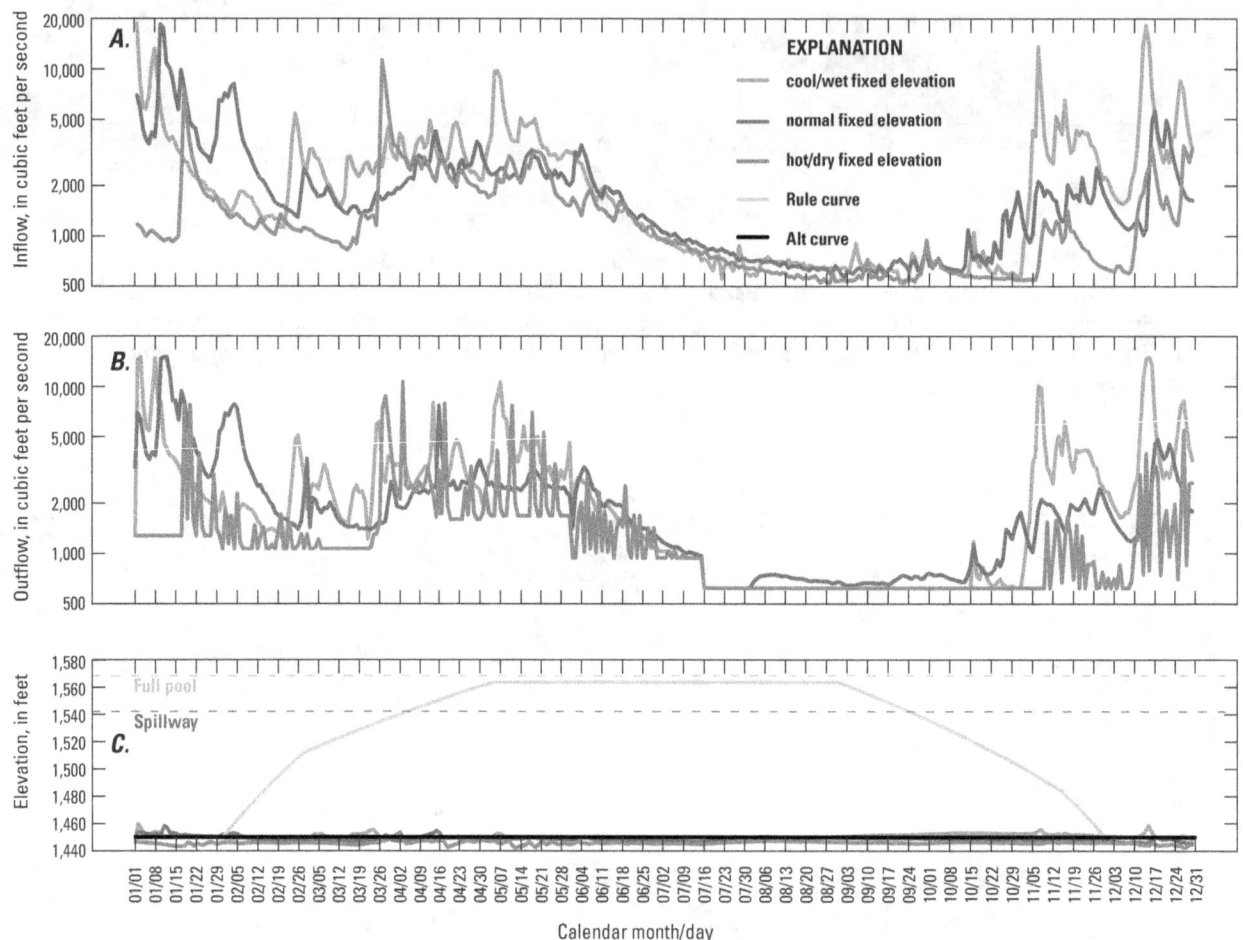

Figure 12.   Comparison of *fixed_elevation* operational scenarios (scenarios *c3, n3, h3*), (*A*) total inflows, (*B*) total outflows, (*C*) modeled water-surface elevation, North Santiam River, Oregon.

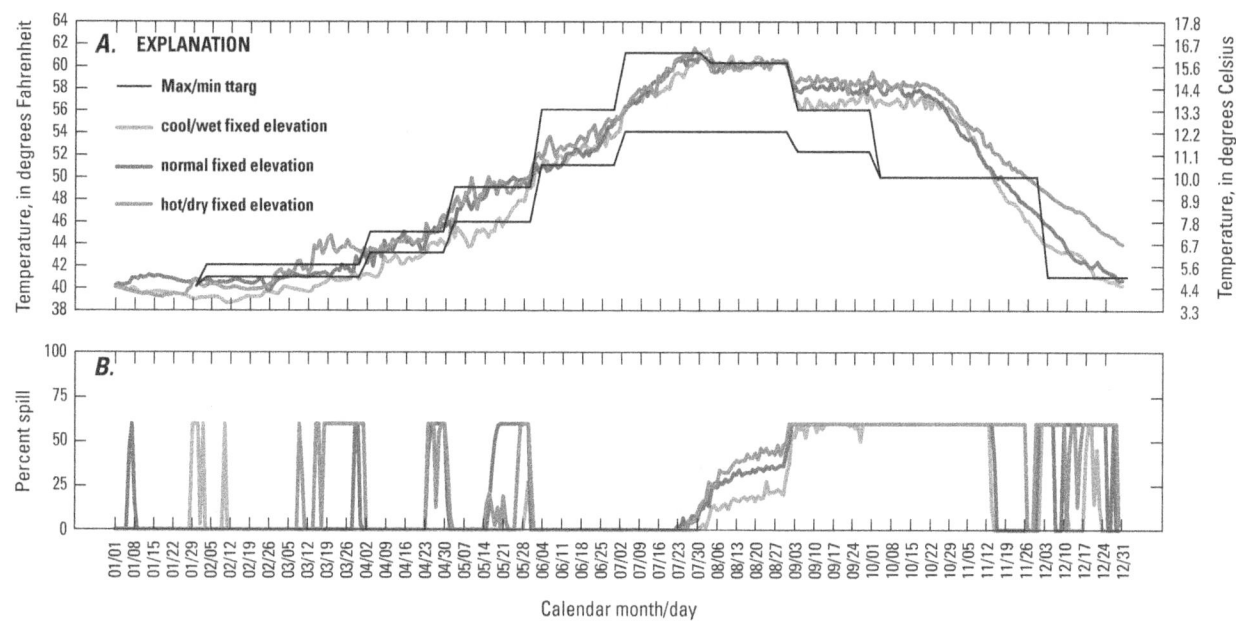

Figure 13. (*A*) Modeled water temperature and (*B*) percent spill for *existing* structural scenarios with *fixed elevation* operational scenarios and *max* temperature targets (scenarios *c3, n3, h3*), North Santiam River, Oregon. The maximum and minimum temperature target established for the McKenzie River (labeled "Max/min ttarg") is shown but only the maximum was used in this simulation.

Delayed Drawdown

Reduced Minimum Outflows

When minimum outflows were decreased in summer under operational scenario *delay_dd1* (tables 4, 5), the lake remained closer to full in mid- and late-summer until the rule curve dictated that the lake be drafted (lowered) to make room for potential flood storage. In this scenario, drawdown typically began in mid- to late-September under all three environmental scenarios (fig. 14), but the rule curve was not modified. Decreased outflows and a higher lake level in midsummer meant that the spillway crest was slightly deeper and accessing slightly cooler water, resulting in cooler releases at that time compared to *base* operations (figs. 8, and 15). In contrast, keeping a higher lake level in early to mid-September meant that the spillway could be used later in the summer compared to *base* operations (figs. 7, and 14), allowing more warm water to be expelled in early September and saving some cooler water for release later in autumn. This generally led to lower outflow temperatures in autumn when compared to *base* operations (figs. 8, and 15).

When reduced minimum outflow operations were adjusted to free the model from the rule directing a minimum 40 percent outflow to the power penstocks (scenario *delay_dd1_noppmin*), the modeled release temperatures from Detroit Dam generally met the *max* temperature target in autumn (fig. 16). Given this scenario allowing the outlets to access both warm water near the lake surface and cool water at depth with few restrictions on minimum flows, downstream temperature targets generally can be met.

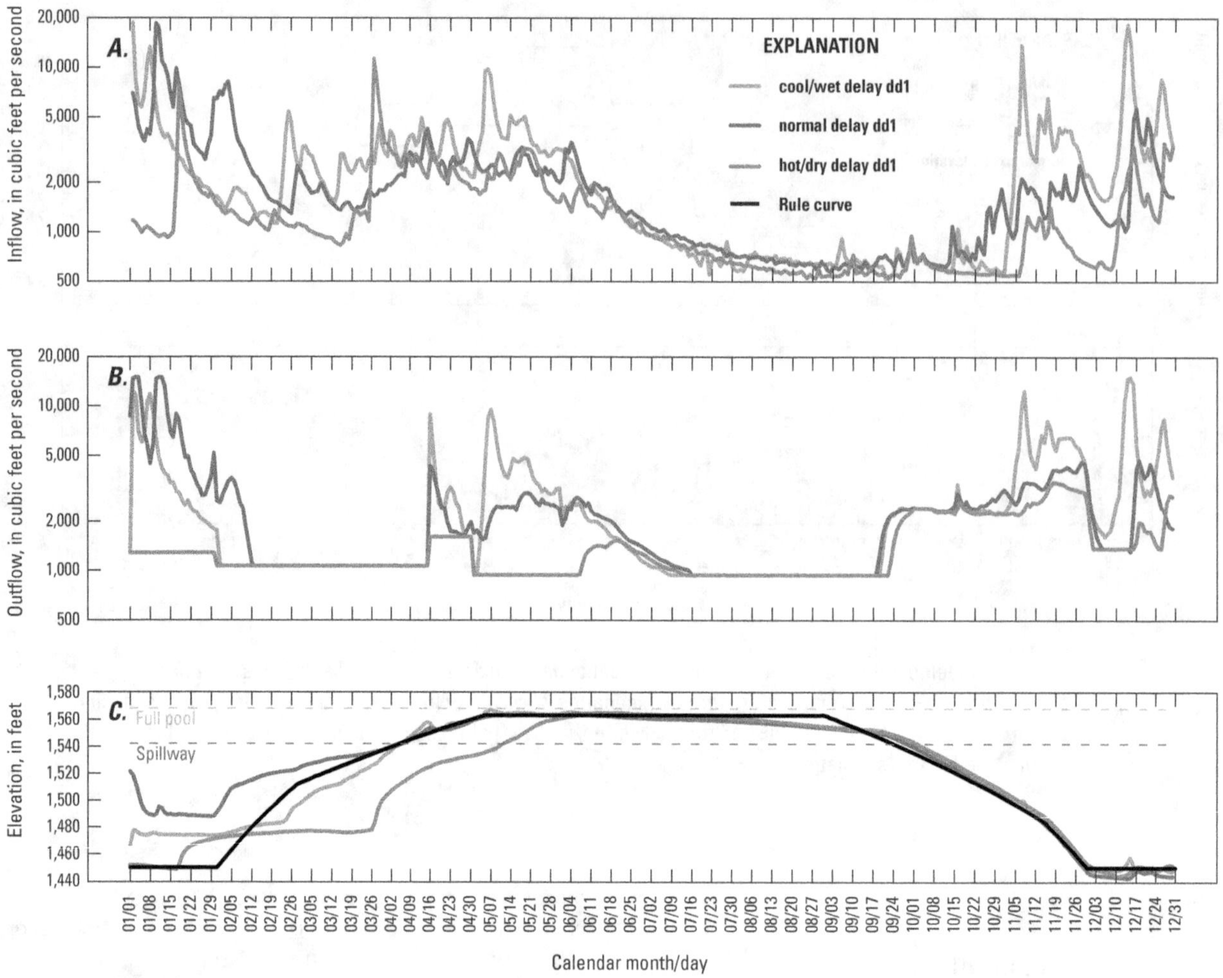

Figure 14. Comparison of *delay_dd1* operational scenarios (scenarios *c6*, *n6*, *h6*), (*A*) total inflows, (*B*) total outflows, (*C*) modeled water-surface elevation, North Santiam River, Oregon.

Reduced Minimum Outflows and Modified Rule Curve

Another way to keep more water in the lake and retain the use of the spillway in late summer is to modify the rule curve, keeping the target lake level higher late in the season and delaying the drawdown that is done to make room for potential flood storage. To examine the effects of delaying drawdown by about two months while minimizing reductions to summertime *base* operational scenario minimum releases, the *"delay_dd2"* operational scenario was developed (tables 4, 5). In this scenario, minimum releases were not decreased in midsummer as in the *delay_dd1* operations; minimum

releases were only modified after September 1. Drawdown under *delay_dd2* typically began in late-October under all environmental scenarios except for *hot/dry* (fig. 17), in which minimum summertime outflow rules did not allow for the lake to remain above the spillway any later than it did under *base* operations. This scenario led to temperature management in autumn that generally was more successful than under *base* operations (compare figs. 8, and 18). During October and November, the *cool/wet* and *normal* scenarios under *delay_dd2* operations resulted in simulated outflow temperatures that generally did not exceed the *max* temperature target (fig. 18). Results were not necessarily better than those using *delay_dd1* operations (fig. 15).

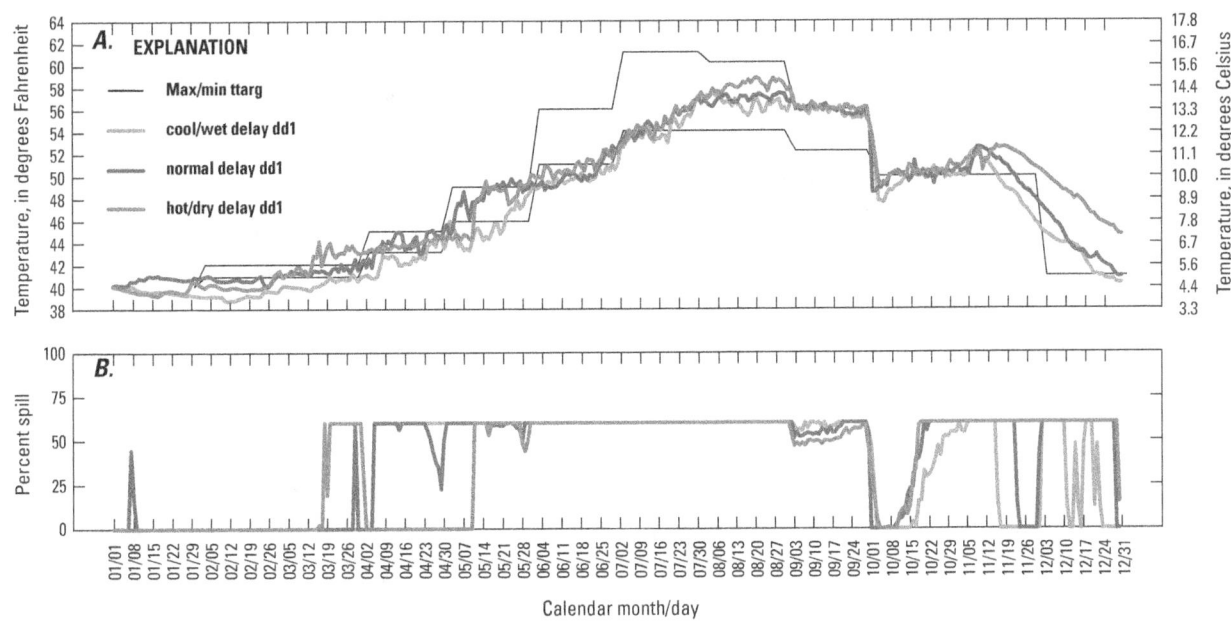

Figure 15. (*A*) Modeled water temperature and (*B*) percent spill for *existing* structural scenarios with *delay_dd1* operational scenarios and *max* temperature targets (scenarios *c6*, *n6*, *h6*), North Santiam River, Oregon. The maximum and minimum temperature target established for the McKenzie River (labeled "Max/min ttarg") is shown but only the maximum was used in this simulation.

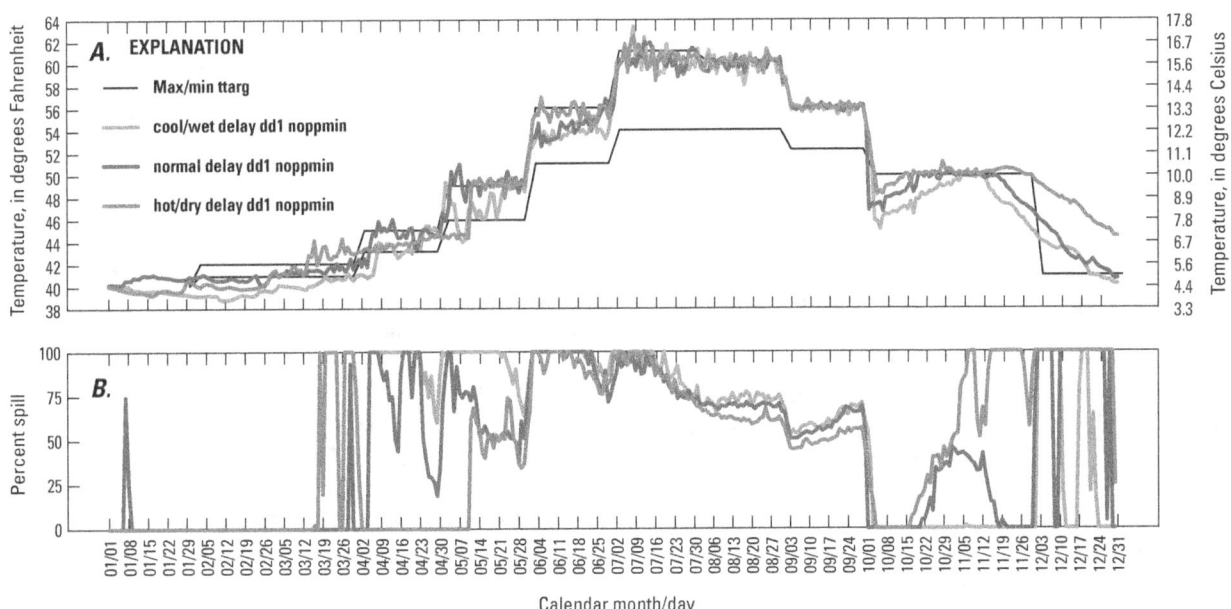

Figure 16. (*A*) Modeled water temperature and (*B*) percent spill for *existing* structural scenarios with *delay_dd1_ noppmin* operations and *max* temperature targets (scenarios *c7*, *n7*, *h7*), North Santiam River, Oregon. The maximum and minimum temperature target established for the McKenzie River (labeled "Max/min ttarg") is shown but only the maximum was used in this simulation.

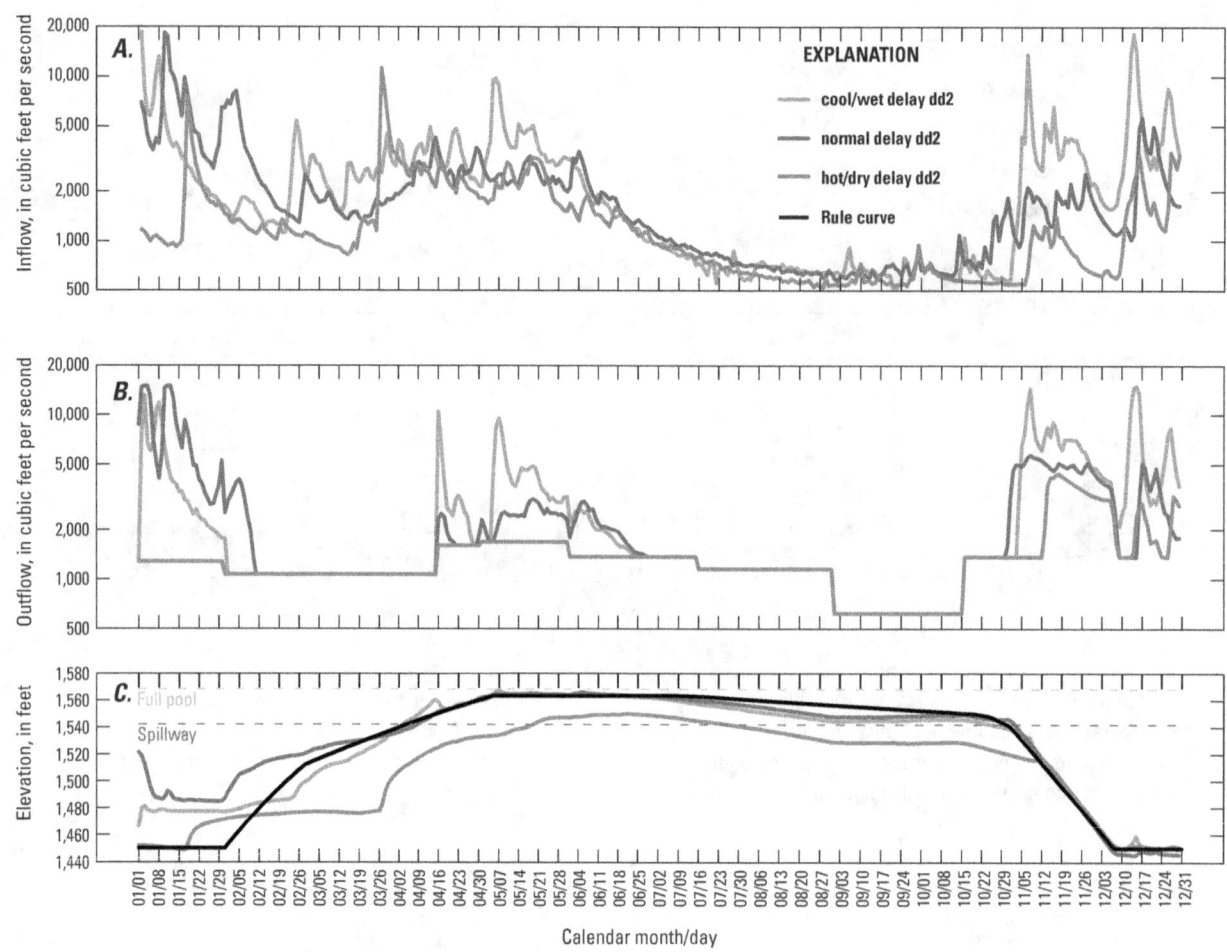

Figure 17.   Comparison of *delay_dd2* operational scenarios (scenarios *8, 9,* and *18*), (*A*) total inflows; (*B*) total outflows, (*C*) modeled water-surface elevation, North Santiam River, Oregon.

Combining the delayed drawdown with no minimum outflow to the power penstocks (*delay_dd2_noppmin* in table 5) resulted in modeled outflow temperatures from Detroit Dam that generally met the *max* temperature target during autumn, aside from a period in November under the *hot/dry* environmental scenario that exceeded the *max* temperature target (fig. 19). As was noted for other scenarios with no minimum power generation, such a scenario allowed for more heat at the lake surface to be discharged during midsummer, thus meeting the *max* temperature target at that time and saving the deeper cold water for release in autumn when cooler releases were required by the temperature targets.

Although delaying the drawdown of the lake may be beneficial for downstream temperature management under certain conditions, such advantages must be balanced against the need to provide protection against potential flood damages. The delayed drawdown simulated in these scenarios specifies a drawdown of the lake that occurs primarily in November, which historically is one of the months with the greatest precipitation, leading to potentially large lake inflows. Clearly, if the drawdown of the lake is delayed, accurate and precise forecasting techniques must be used to balance the need for temperature management against impending needs to make room in the reservoir to capture high-inflow events.

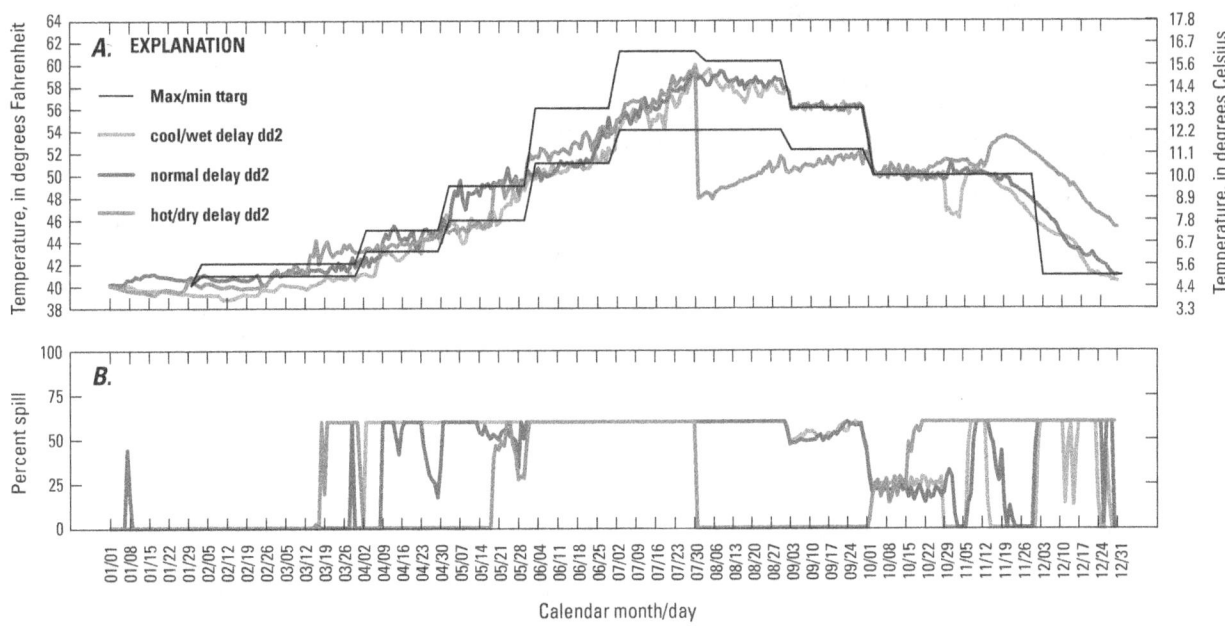

Figure 18. (*A*) Modeled water temperature and (*B*) percent spill for *existing* structural scenarios with *delay_dd2* operations and *max* temperature targets (scenarios *c8*, *n8*, *h8*), North Santiam River, Oregon. The maximum and minimum temperature target established for the McKenzie River (labeled "Max/min ttarg") is shown but only the maximum was used in this simulation.

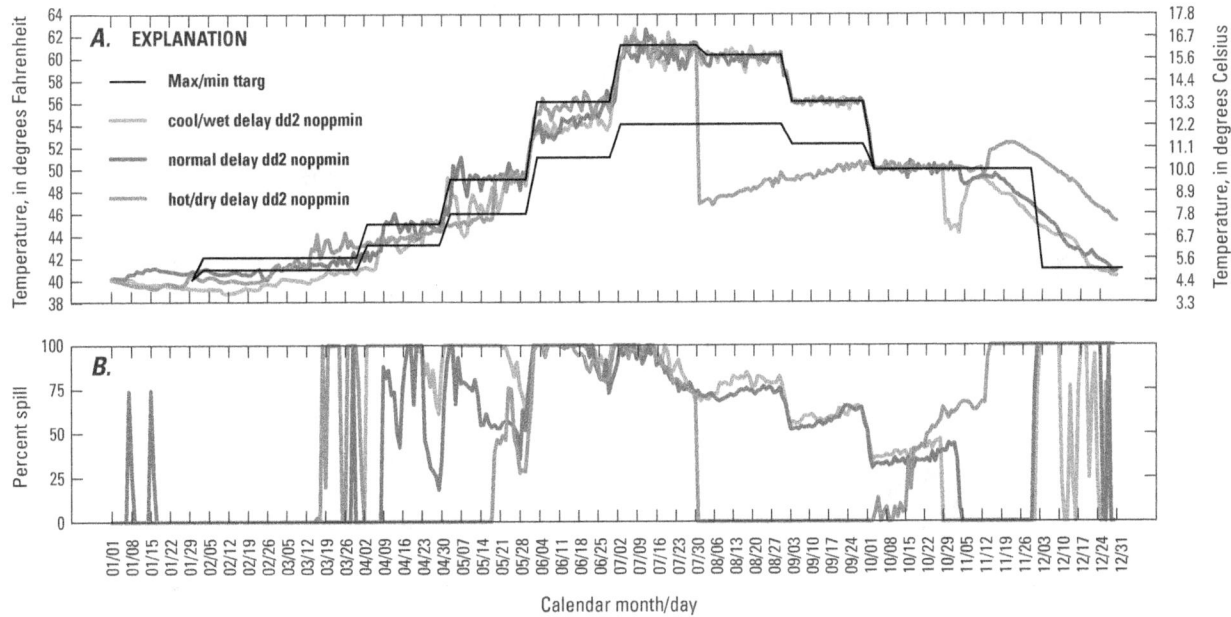

Figure 19. (*A*) Modeled water temperature and (*B*) percent spill for *existing* structural scenarios with *delay_dd2_ noppmin* operations and *max* temperature targets (scenarios *c9*, *n9*, *h9*), North Santiam River, Oregon. The maximum and minimum temperature target established for the McKenzie River (labeled "Max/min ttarg") is shown but only the maximum was used in this simulation.

Detroit Dam Structural Scenarios

Structural scenarios using the Detroit Lake model were limited only by the three possible types of outlets that are available in the USGS-coded CE-QUAL-W2 v3.12 blending routine: fixed-elevation, floating, or sliding-gate (variable-elevation). Sliding-gate outlets have user-specified vertical limits in the depth of the lake, whereas floating outlets are located at a user-defined depth below the water surface, and have a lower vertical limit. Similarly, sliding-gate outlets are positioned at a user-defined depth below the lake surface when the blending routine calls for such an outlet to be positioned at or near the lake surface. For this study, a lower vertical limit of 1,340 ft (the elevation of the upper ROs) and a depth of 6.6 ft (2 m) below the lake surface were specified for all floating and sliding-gate outlets. Four possible combinations of fixed-elevation, floating, and sliding-gate outlets, as well as the existing fixed-elevation outlets, were used in separate groups of structural scenarios in this study (table 6).

Single Sliding-Gate Structure

Replacing all outlets at Detroit Dam with a single sliding-gate assembly simplified the modeled operations greatly. No minimum power generation constraints were set because presumably any or all of the flow through the new outlet could be routed to the hydropower plant. The minimum and maximum outflow rates of the *base* operations still applied. This structural scenario was named *slider1340* because the lower elevation limit was set at 1,340 ft, the elevation of the upper ROs.

Scenarios in which a single sliding-gate outlet was used led to modeled outflow temperatures that generally varied more on a daily basis compared to scenarios using more than one outlet (fig. 20). This tendency was especially evident in autumn. The large variation in release temperatures is a result of the sliding-gate outlet being positioned at a depth that often was located in or near the middle of the thermocline, such that any seiching of the lake caused the thermocline to move up and down over the course of the day and thereby change the temperature of the water captured by the outlet. The model scenario was configured so that the elevation of the sliding-gate outlet was adjusted by the model only once per day (at 0500 hours), in order to minimize demands on dam operators; similar criteria were used for other scenarios such that gate adjustments were generally only performed once a day. Despite the larger daily variations in release temperatures, use of a single sliding-gate outlet in these *slider1340* scenarios generally allowed downstream temperature targets to be met, with the exception of occasional spikes in autumn and general exceedances during December (fig. 20).

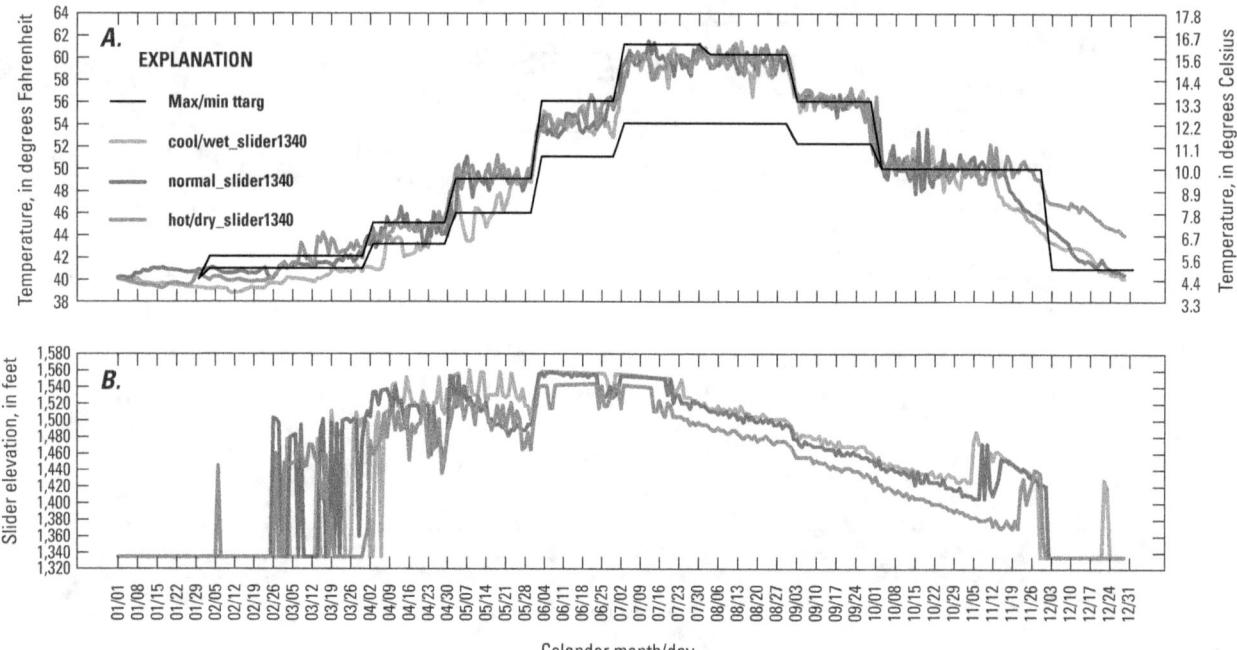

Figure 20.   (*A*) Modeled water temperature and (*B*) sliding-gate elevation for *slider1340* structural scenarios with *base* operations and *max* temperature targets (scenarios *c17*, *n17*, *h17*), North Santiam River, Oregon. The maximum and minimum temperature target established for the McKenzie River (labeled "Max/min ttarg") is shown but only the maximum was used in this simulation.

Floating and Fixed-Elevation Gates

Structural scenarios in which a fixed-elevation outlet and a floating outlet were used in combination (*pp-float* and *uro-float*) led to modeled release temperatures that were similar to results from scenarios with a single sliding outlet (*slider1340*), but the former generally displayed less daily variation than the latter. The *pp-float* and *uro-float* scenarios depict the existing power penstocks and upper RO gates, respectively, as fixed-elevation outlets used in combination with a new floating outlet.

Floating Outlet with Power Penstocks

The *pp-float* structural scenarios specify one hypothetical floating outlet 6.6 ft (2 m) below the lake surface as well as a fixed outlet at the elevation of the existing power penstocks (1,402.9 ft centerline elevation). A series of operational scenarios were modeled in combination with these outlets to determine the effect of different minimum flow requirements to the power penstocks and the floating outlet. Results showed that as the minimum of the total outflow directed to the fixed-elevation (power) outlet was increased, the release temperatures generally decreased in spring and increased in the autumn. This effect is visible in figures 21, 22, 23, and E5 by comparing the discharge rate from the fixed outlet and the resulting outflow temperature in each scenario.

No Minimum Flow through Power Penstocks or Floating Outlet

With no minimum flow directed to either outlet, the model was free to optimize the release temperatures based on the lake temperatures near each outlet. As a result, the release temperatures generally met the *max* temperature target in scenarios *c10* and *n10*. The *hot/dry* scenario (*h10*), however, resulted in an undesirable peak outflow temperature of about 54°F in autumn (fig. 21).

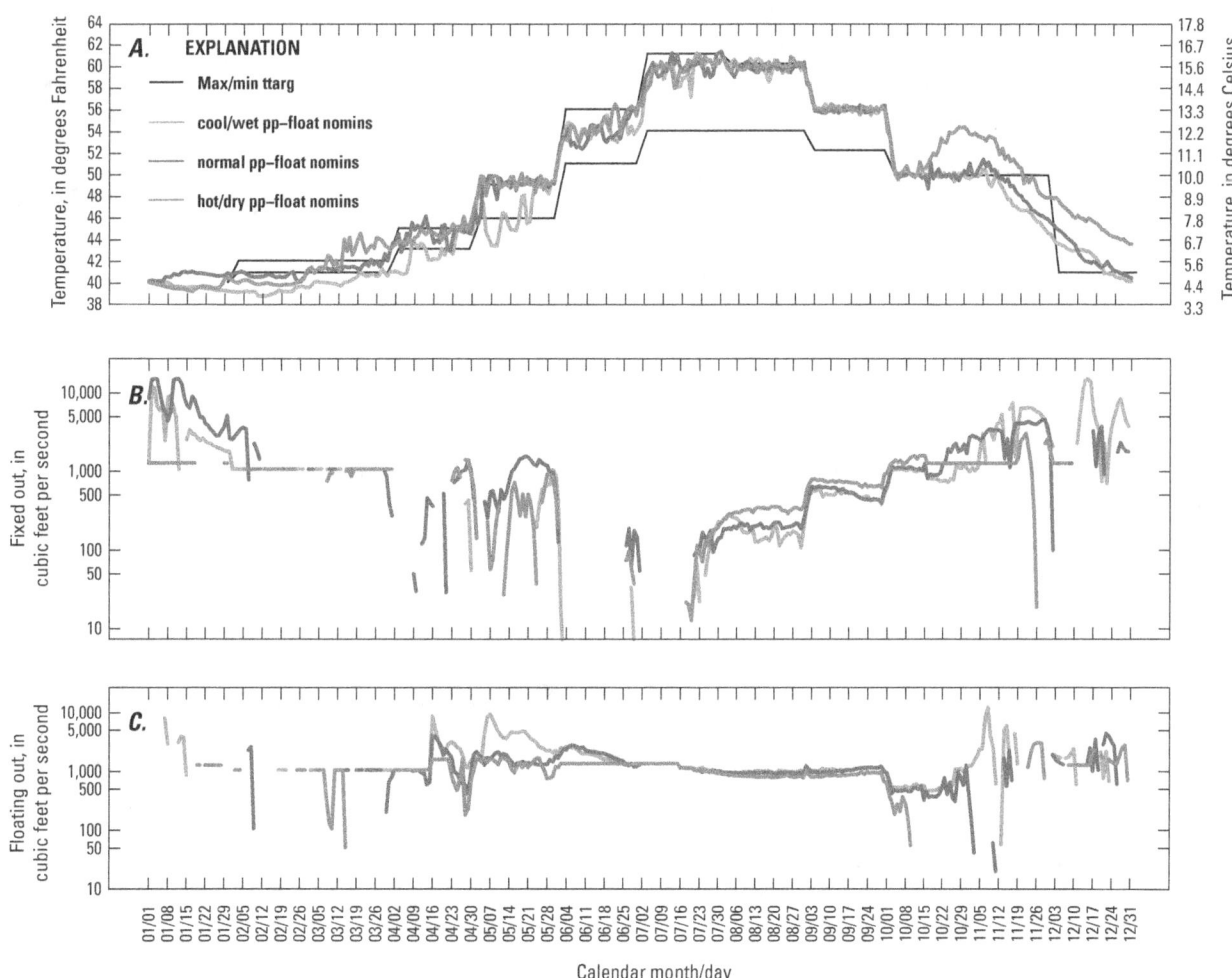

Figure 21. Results from *pp-float* structural scenarios with *nomins* operations and *max* temperature targets (scenarios *c10*, *n10*, *h10*), (*A*) modeled water temperature, (*B*) outflow from fixed outlet, and (*C*) outflow from floating outlet, North Santiam River, Oregon. The maximum and minimum temperature target established for the McKenzie River (labeled "Max/min ttarg") is shown but only the maximum was used in this simulation.

Twenty Percent Minimum Flow through Power Penstocks

As the minimum outflow to the power penstocks is increased from 0 to 20 percent (scenarios *c12, n12,* and *h12* in table 7), the amount of warm surface water that can be released via the floating outlet is decreased, resulting in cooler releases in June and July and warmer releases during autumn (fig. 22). The warmer releases in autumn are a direct result of releasing more of the deeper, cooler water in midsummer, which decreases the reserves of cool water at the level of the power penstocks and draws the thermocline down to deeper depths, thus pulling warmer water to the elevation of the power penstocks in autumn.

Minimum Flow of 400 Cubic Feet per Second through Floating Outlet

The use of a floating outlet has two potential benefits. First, it allows continual access to warm water at the top of the lake in spring through autumn. Second, it can provide a means of collecting fish for downstream passage. Certain engineering design criteria and the use of the floating outlet for fish passage might require that the outlet be operated with a minimum flow rate. By placing a 400-ft^3/s minimum outflow requirement on the floating outlet (fig. 23), the release temperatures are quite similar to those that result when no constraints are placed on either outlet (fig. 21). As with the

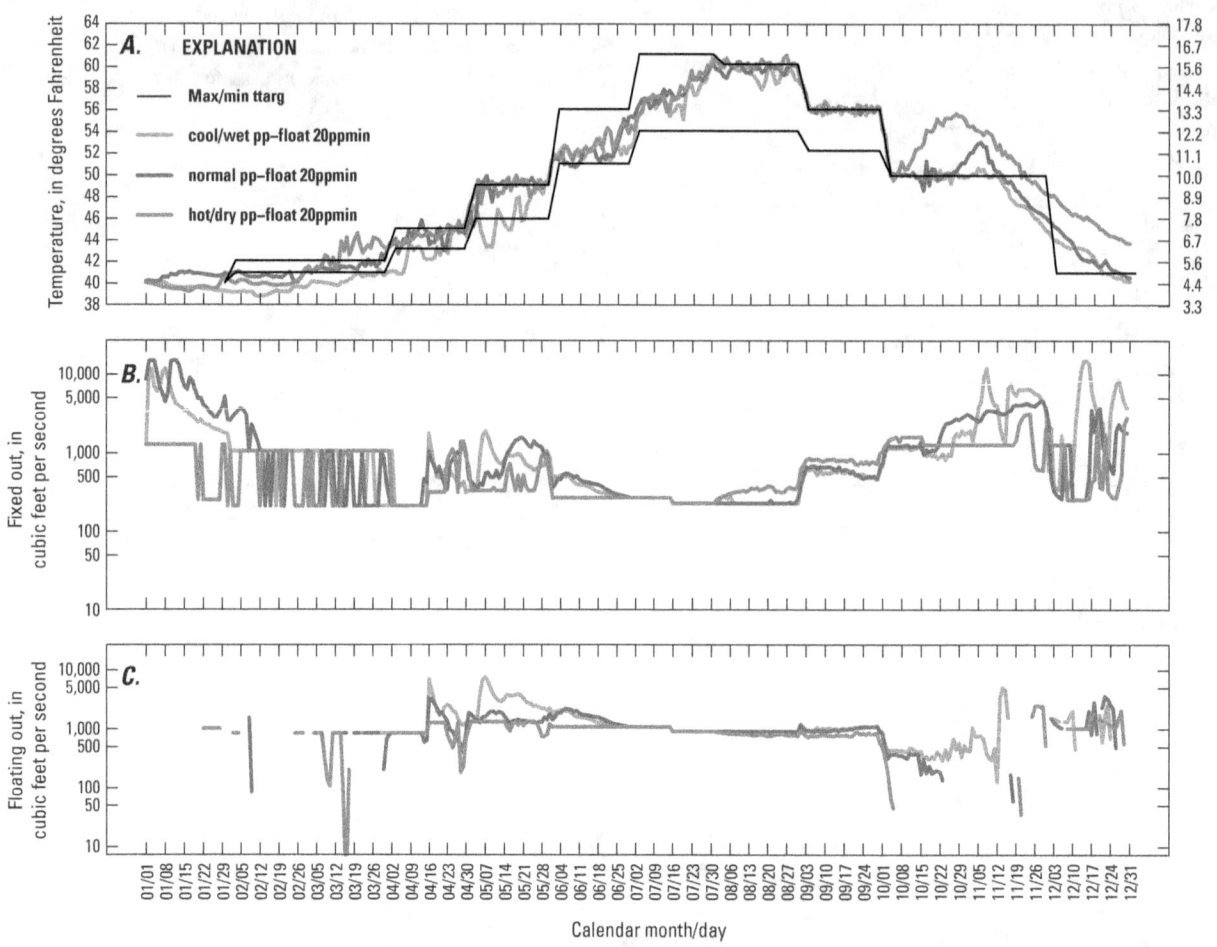

Figure 22. Results from *pp-float* structural scenarios with *20ppmin* operations and *max* temperature targets (scenarios *c12, n12, h12*), (*A*) modeled water temperature, (*B*) outflow from fixed outlet, and (*C*) outflow from floating outlet, North Santiam River, Oregon. The maximum and minimum temperature target established for the McKenzie River (labeled "Max/min ttarg") is shown but only the maximum was used in this simulation.

pp-float scenario that required no minimum flows through either outlet, this scenario exceeds the target temperatures in autumn for the *hot/dry* environmental scenario, mainly because the lake level is low under those conditions and neither outlet is able to access water that is deep enough to still be cool at that time of year.

Floating Outlet with Upper Regulating Outlets

The *uro-float* structural scenarios specify one hypothetical floating outlet 6.6 ft (2 m) below the lake surface as well as a fixed outlet at the elevation of the existing upper ROs (centerline elevation of 1,339.9 ft). These scenarios were developed under the assumption that outflow from upper ROs could be routed to the powerhouse at Detroit Dam for power production; therefore, the same sort of operational scenarios for a minimum amount of power generation were applied.

Similar to results from the *pp-float* structural scenarios, as the outflow directed to the fixed-elevation outlet was increased, the outflow temperatures generally decreased in spring and increased in autumn. This trend can be seen in figures 24, 25, and 26 by comparing the discharge from the fixed-elevation outlet (labeled "Fixed out") and the outflow temperature in each scenario. Greater releases of cool water from depth in midsummer generally diminish the probability of meeting June and July temperature targets and deplete the reservoir of cool water that is available for release in autumn.

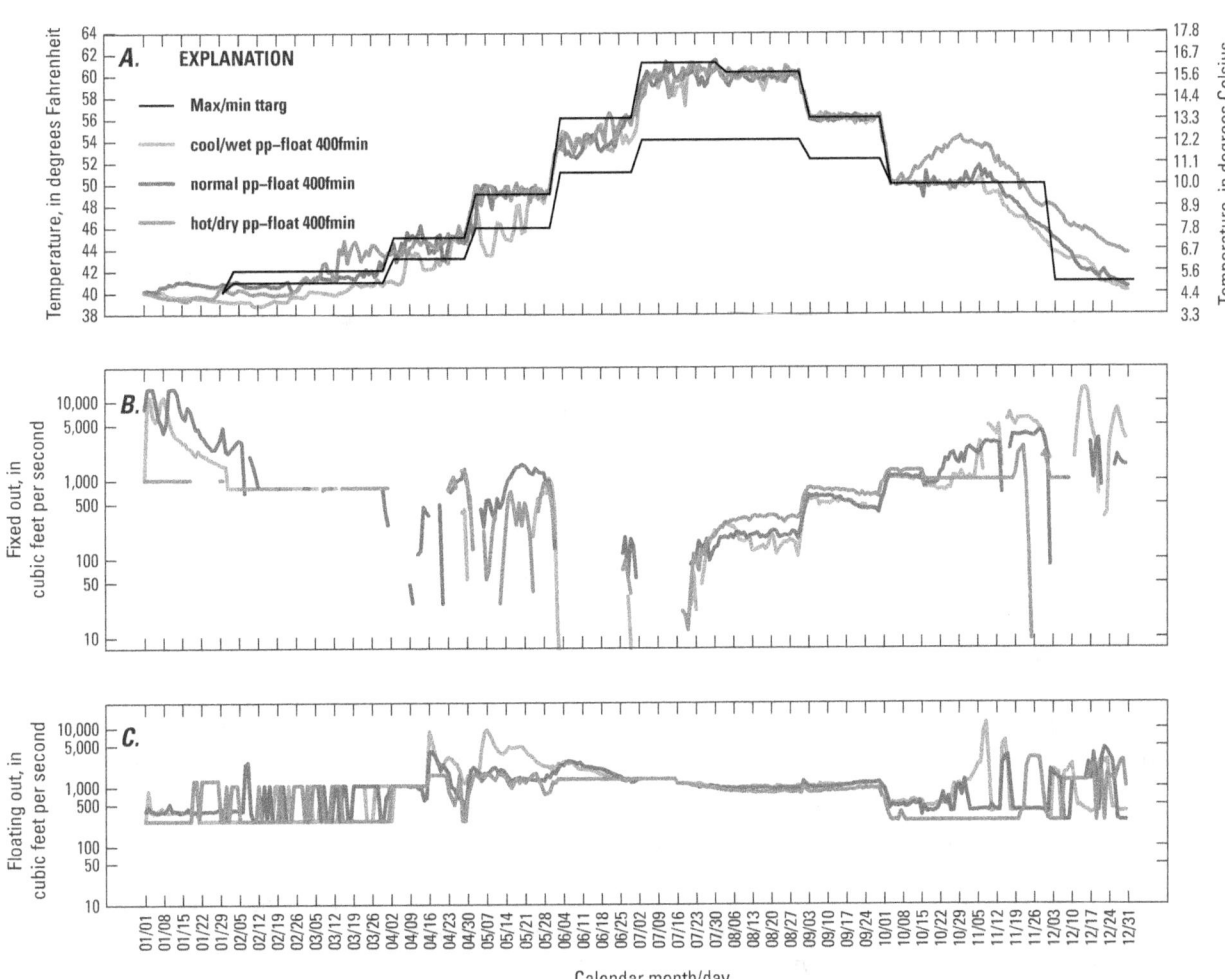

Figure 23. Results from *pp-float* structural scenarios with *400fmin* operations and *max* temperature targets (scenarios *c13, n13, h13*), (*A*) modeled water temperature, (*B*) outflow from fixed outlet, and (*C*) outflow from floating outlet, North Santiam River, Oregon. The maximum and minimum temperature target established for the McKenzie River (labeled "Max/min ttarg") is shown but only the maximum was used in this simulation.

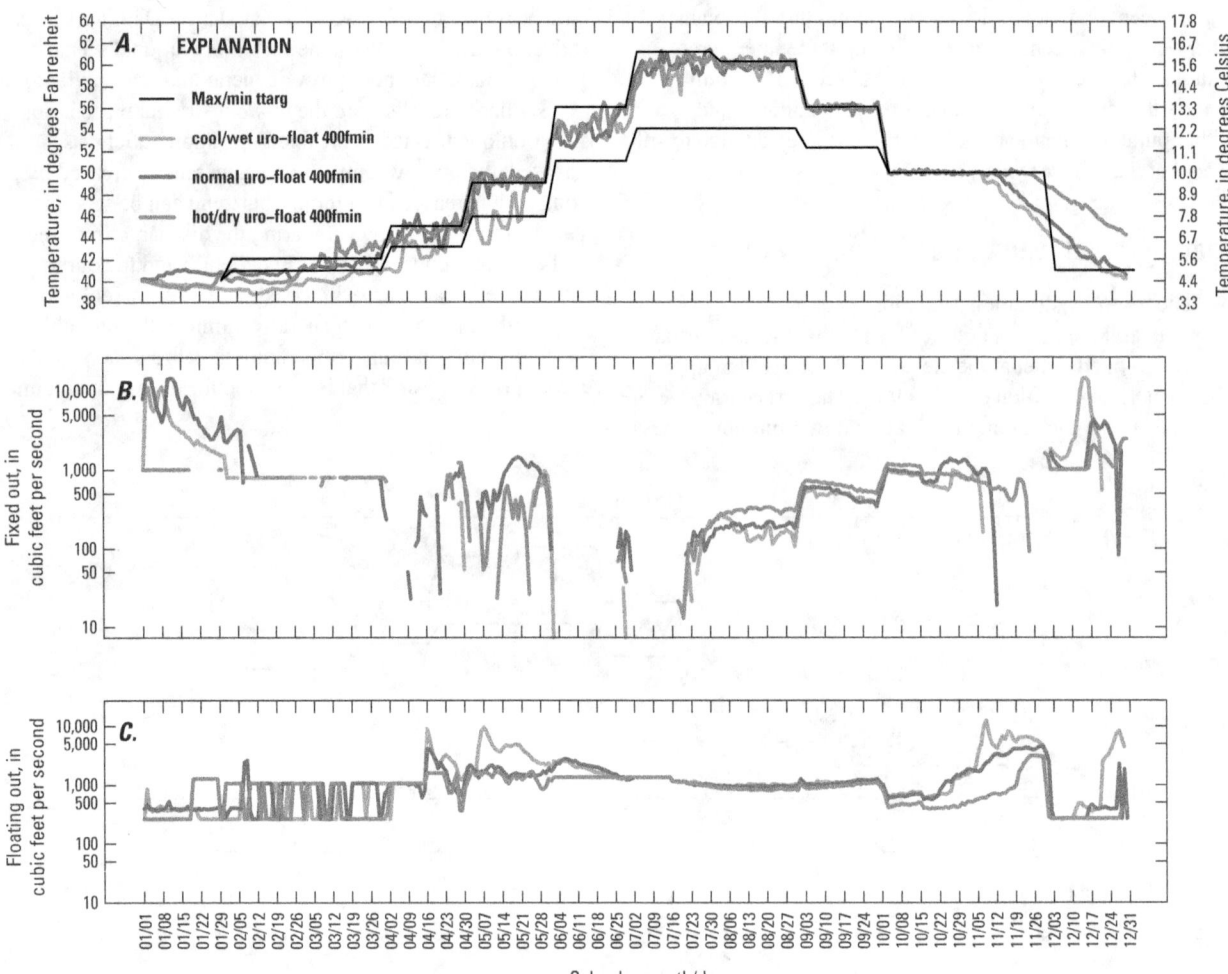

Figure 24. Results from *uro-float* structural scenarios with *400fmin* operations and *max* temperature targets (scenarios *c14*, *n14*, *h14*), (*A*) modeled water temperature, (*B*) outflow from fixed outlet, and (*C*) outflow from floating outlet, North Santiam River, Oregon. The maximum and minimum temperature target established for the McKenzie River (labeled "Max/min ttarg") is shown but only the maximum was used in this simulation.

Minimum Flow of 400 Cubic Feet per Second through Floating Outlet

One potential floating withdrawal structure design for Detroit Dam would have the ability to convey fish downstream given a minimum flow requirement through the structure. When a 400-ft³/s minimum outflow requirement is placed on the floating outlet and no minimum flow is directed to the upper ROs, the result is that the vast majority of outflow is directed to the floating outlet during June–July to meet the *max* temperature target. This leads to cooler outflow temperatures in autumn, when scenarios *c14*, *n14*, and *h14* generally met the temperature target (fig. 24). Exporting more heat from the lake surface in midsummer allows dam operators to reserve more of the cool water at depth for use in autumn. Decreased export of water from depth in midsummer

also means that the thermocline is not drawn down as far, helping to retain access to cool water below the thermocline at the fixed-elevation outlet in autumn.

Twenty Percent Minimum Flow through Upper ROs

Requiring at least 20 percent of the total outflow to pass through the upper ROs in scenarios *c15*, *n15*, and *h15* results in releases from the ROs that do not fall below about 250 ft³/s during summer months. Under the *uro-float_20ppmin* and *uro-float_40ppmin* scenarios, it was assumed that power production could be routed through the upper ROs. These scenarios generally result in outflow temperatures close to the *max* temperature target during autumn (fig. 25), showing that some minimum amount of power can be generated while still meeting downstream temperature targets.

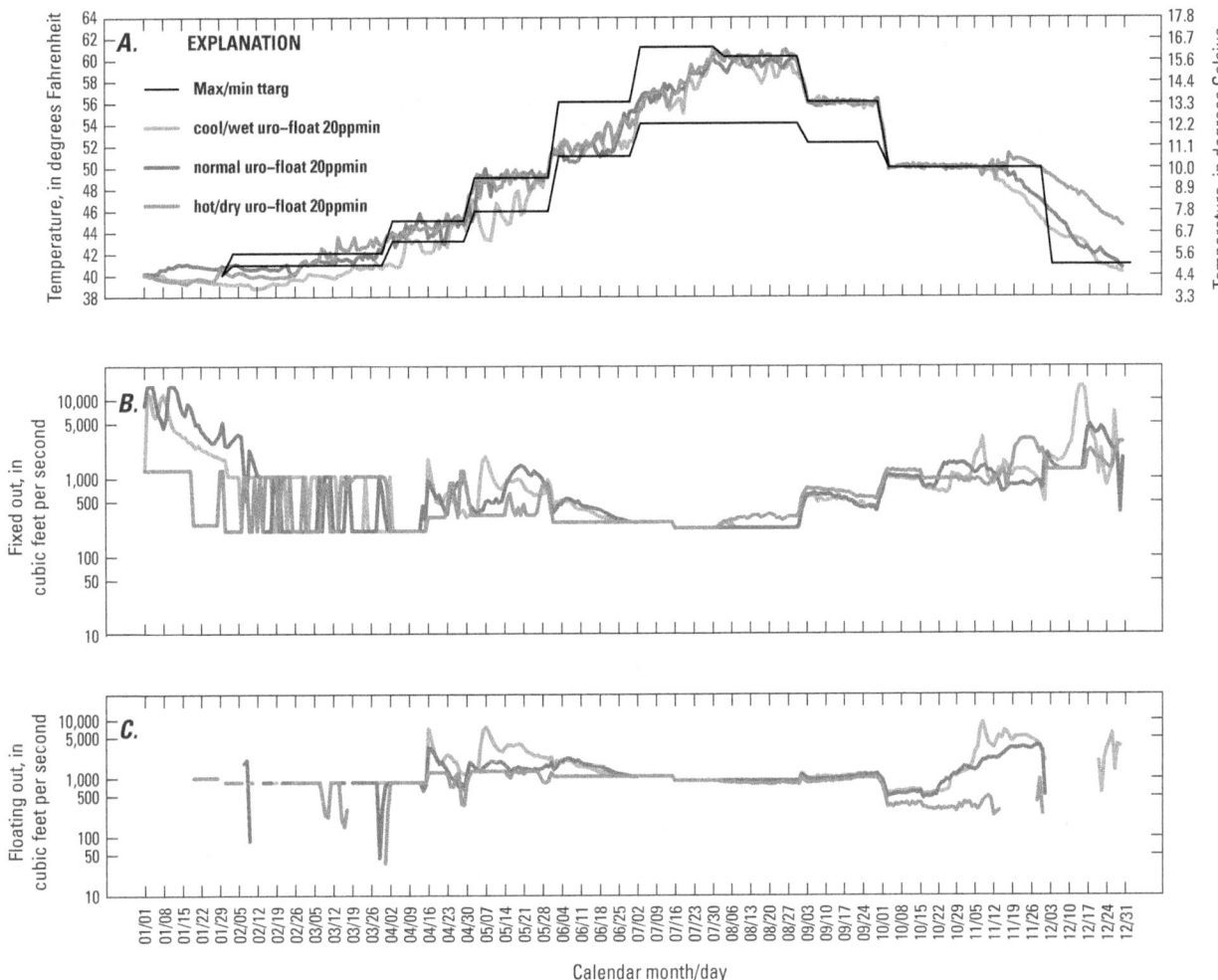

Figure 25. Results from *uro-float* structural scenarios with *20ppmin* operations and *max* temperature targets (scenarios *c15, n15, h15*), (*A*) modeled water temperature, (*B*) outflow from fixed outlet, and (*C*) outflow from floating outlet, North Santiam River, Oregon. The maximum and minimum temperature target established for the McKenzie River (labeled "Max/min ttarg") is shown but only the maximum was used in this simulation.

Forty Percent Minimum Flow through Upper ROs

By increasing the minimum outflow requirement on the upper RO gates to 40 percent, outflow from the upper ROs does not fall below about 500 ft³/s during summer months (fig. 26B). These scenarios generally result in outflow temperatures that exceed the *max* temperature target during November (fig. 26A). Clearly, as more water is drawn from below the thermocline in midsummer, less of the cool water below the thermocline is accessible in autumn.

Sliding and Floating Gates

Structural scenarios using a combination of a sliding-gate and a floating outlet were run to evaluate how access to both warm surface water and cool water at depth would allow downstream temperature targets to be met under a range of conditions. The sliding-gate outlet was assigned a lower vertical limit of 1,340 ft, which is similar to the elevation of the upper ROs. Similar to scenarios depicting a single sliding-gate outlet, these *slider1340-float* structural scenarios resulted in outflow temperatures near the *max* temperature target for most of the calendar year except for the month of December.

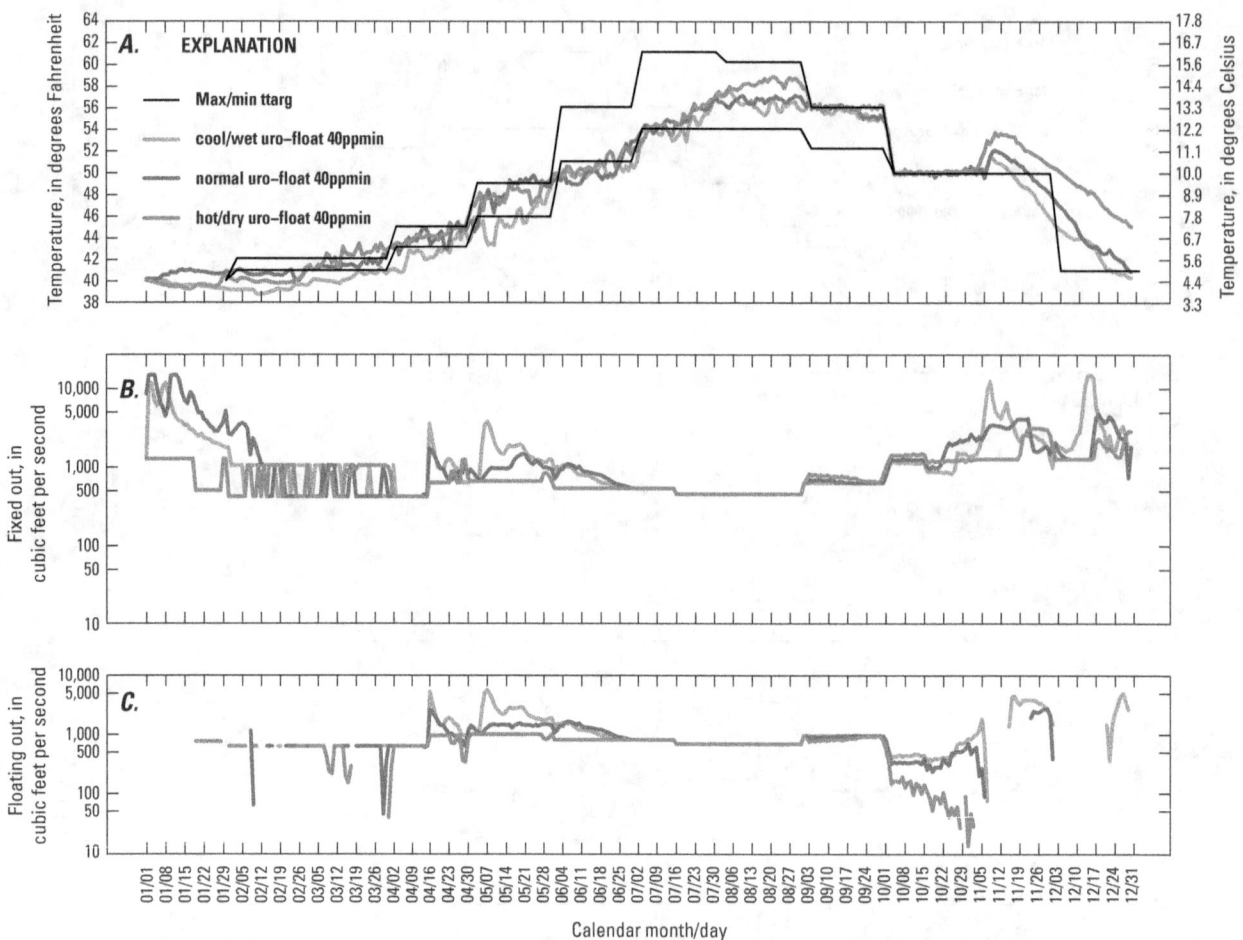

Figure 26.   Results from *uro-float* structural scenarios with *40ppmin* operations and *max* temperature targets (scenarios *c16*, *n16*, *h16*), (*A*) modeled water temperature, (*B*) outflow from fixed outlet, and (*C*) outflow from floating outlet, North Santiam River, Oregon. The minimum temperature target established for the McKenzie River is shown but not used in this simulation.

Fixed Flow of 400 Cubic Feet per Second through Floating Outlet

By simulating a constant flow of 400 ft³/s to the floating outlet, this scenario was designed to represent the potential effects of a hypothetical year-round lake-surface withdrawal structure that might also accommodate fish passage. Results from scenarios *c19*, *n19*, and *h19* show that *max* temperature targets generally could be met throughout the year with this outlet configuration (fig. 27), although the temperature target was actually exceeded at times in August–September due to

the large surface outflow. The USGS-coded CE-QUAL-W2 v3.12 blending routine does not explicitly solve for the mixed temperature between two outlets when a constant flow is designated to one outlet, so some of the exceedances in outflow temperature during late August and early October (fig. 20A) may be due to inconsistencies between the imposed temperature target and the blended outflow temperature calculated by the model. A modified blending subroutine could fix this problem, but the point is that this scenario can come close to meeting the temperature target most of the time.

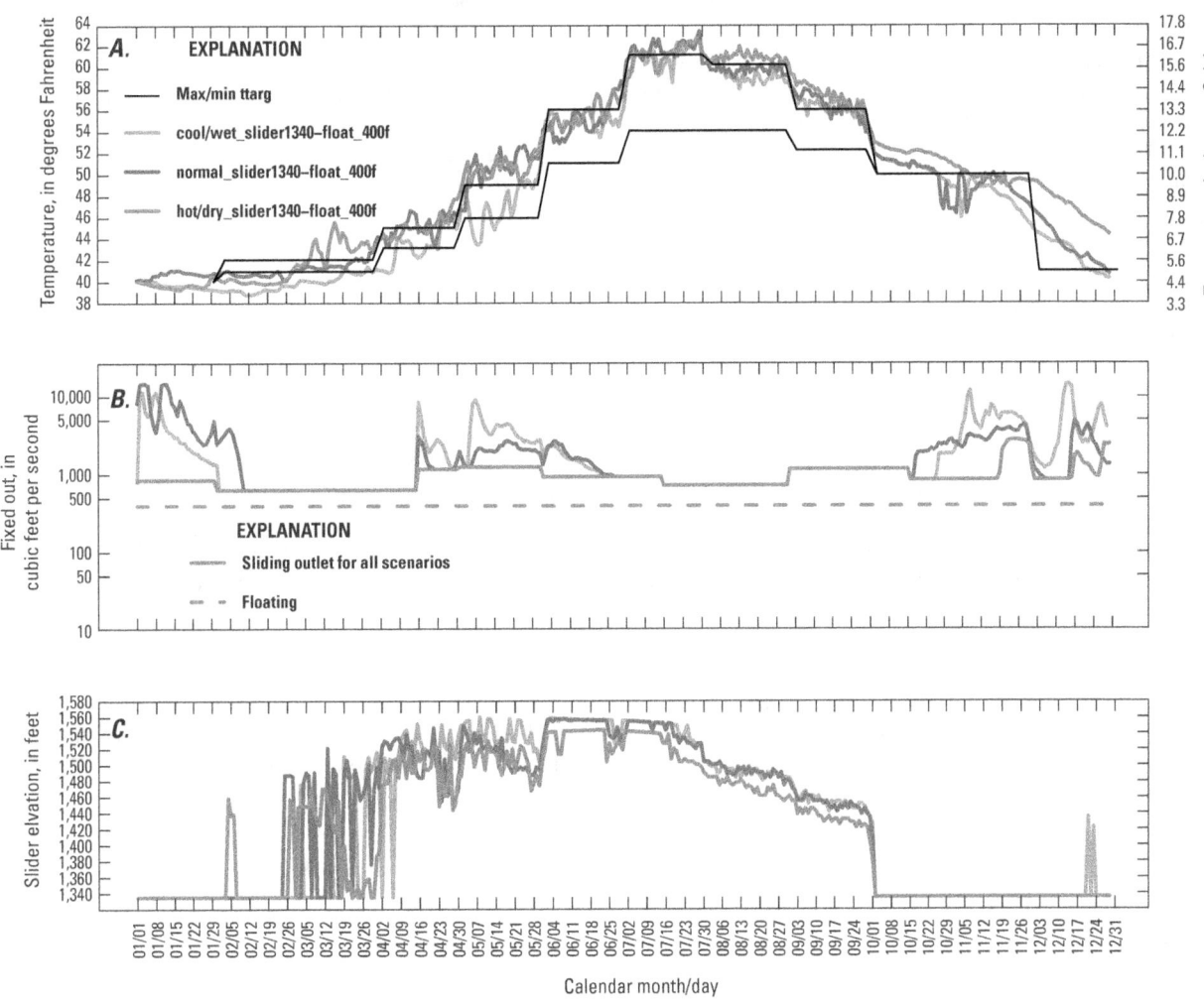

Figure 27. Results from *slider1340-float* structural scenarios with *400f* operations and *max* temperature targets (scenarios *c19*, *n19*, *h19*), (*A*) modeled water temperature, (*B*) outflow from fixed outlet, and (*C*) sliding outlet elevation, North Santiam River, Oregon. The maximum and minimum temperature target established for the McKenzie River (labeled "Max/min ttarg") is shown but only the maximum was used in this simulation.

Downstream Effects of Selected Scenarios

To assess the potential downstream effects of operational and structural changes at Detroit Dam, the Big Cliff Reservoir and North Santiam-Santiam River models were used to run a subset of the Detroit Dam scenarios. Not all scenarios were run through the downstream models because the downstream patterns of temperature change are likely to be similar. The selected scenarios were chosen because of their potential for being adopted by USACE as they evaluate possible operational and/or structural changes to Detroit Dam.

Estimated Emergence Dates

The Accumulated Thermal Unit (ATU) is a "degree-day" type of calculation used to estimate the date on which spring Chinook salmon first emerge from their eggs (U.S. Army Corps of Engineers, 2012). The ATU calculation in this report is the cumulative sum of the daily average temperature (in degrees Fahrenheit) exceeding 32°F beginning at the typical peak of spring Chinook spawning on September 20. The estimated emergence day then is derived as the date when the ATU value reaches 1,750°F-day. These emergence day estimates are based on observed egg emergences at the Oregon Department of Fish and Wildlife Willamette Hatchery in Oakridge, Oregon, when the ATU value is 1,650–1,850°F-day (U.S. Army Corps of Engineers, 2012).

Simulated Detroit Dam release temperatures were used for this computation. Some model scenarios resulted in estimated emergence dates that were later than December 31. In these cases, the simulated Detroit Dam outflow temperature from each respective scenario during the previous January and February were used to complete the ATU calculation. Estimated emergence dates under each Detroit Lake model scenario are compared and sorted according to the *hot/dry* environmental scenario results in table 8. The *hot/dry* scenarios display a somewhat worst-case potential effect that any Detroit Dam operational/structural scenario might have on downstream water temperatures during a low-flow year.

Scenarios having only operational changes at Detroit Dam resulted in earlier emergence dates than did scenarios with structural changes or structural and operational changes combined, suggesting that some structural changes might be more beneficial for fish under *hot/dry* conditions. Operational scenarios with lower minimum power production requirements, which generally provided cooler release temperatures in autumn, resulted in later estimated emergence dates for each otherwise equivalent scenario. For example, scenario *h2* (*existing* structures, *noppmin* operations, *hot/dry* conditions) resulted in an emergence date 5 days later than *h1* (*existing* structures, *base* operations, *hot/dry* conditions) (table 8).

Emergence date estimates for many of the operational scenarios (*c1, n1, c2, n2, c5, n5, c6, n6, c7, n7*) under *cool/wet* and *normal* environmental scenarios do not follow the same pattern as that exhibited by the *hot/dry* environmental scenarios; instead, the dates are later than those from many of the structural scenarios. This result is primarily caused by outflow temperatures that were well below the temperature target during early to mid-October, a time when the water-surface elevation was below the spillway crest and releases were limited to cooler water from the upper ROs and the power penstocks. The ATU computation has the advantage of integrating all of these conditions into a single numeric value, but it does not convey the rate at which the emergence date is approached or any sequence of events that changes that rate during autumn. Those changes in ATU over time are illustrated in figure 28, showing that many of the scenarios are likely to have quite different effects on the rate of egg incubation during autumn months (fig. 28).

Downstream of Big Cliff Dam in the North Santiam River, estimated emergence dates were calculated from model results (scenarios *h1, h8, h10, h17,* and *h19*) at the location of the USGS gaging station at Mehama (site 14183000, RM 38.7, table 9). All emergence date estimates at Mehama range from 15 to 43 days later than those calculated at the outlet of Detroit Dam. The cooler water temperatures at that site are caused partly by cool inflows from the Little North Santiam River just upstream of Mehama in November–December.

Table 8.   Calculated emergence day for each Detroit Lake model scenario based on the day at which the Accumulated Thermal Units (ATUs) for simulated release temperatures reached 1,750 degrees Fahrenheit-day, North Santiam River, Oregon.

[ATU calculated by the difference of the average daily temperature above 32 degrees Fahrenheit from September 20 through December 31. Dates in January and February were estimated based on model results from each environmental scenario and placed in increasing order according to the *hot/dry* scenarios. **Bold** text refers to scenarios with North Santiam and Santiam River model simulations. "7d_mmwod" prefix signifies the 7-day moving maximum of the without-dams temperature target. All other scenarios were based on the tmax temperature target]

Scenario identifiers	Scenario description	Structural scenario?	Estimated emergence day			Rank			
			cool/wet	normal	hot/dry	cool/wet	normal	hot/dry	Average
c3,n3,h3	_fixed_elevation	No	Dec. 19	Dec. 24	Dec. 5	20	20	20.5	20.2
c4,n4,h4	_late_refill	No	Dec. 18	Dec. 13	Dec. 5	21	21	20.5	20.8
c5,n5,h5	_early_dd	No	Jan. 14	Jan. 5	Dec. 11	19	19	19	19.0
cwod1,nwod1,hwod1	_wod_base	No	Jan. 26	Jan. 13	Dec. 13	4	9.5	17.5	10.3
c1, n1, h1	**_base**	**No**	**Jan. 24**	**Jan. 12**	**Dec. 13**	**5**	**13**	**17.5**	**11.8**
cwod2, nwod2, hwod2	_wod_noppmin	No	Feb. 1	Jan. 18	Dec. 17	2	2	16	6.7
c2, n2, h2	_noppmin	No	Feb. 8	Jan. 22	Dec. 18	1	1	14.5	5.5
c12, n12, h12	_pp-float_20ppmin	Yes	Jan. 19	Jan. 12	Dec. 18	12.5	13	14.5	13.3
c11, n11, h11	_pp-float_10ppmin	Yes	Jan. 21	Jan. 14	Dec. 21	11	8	12.5	10.5
c16, n16, h16	_uro-float_40ppmin	Yes	Jan. 15	Jan. 6	Dec. 21	17.5	17.5	12.5	15.8
c8, n8, h8	**_delay_dd2**	**No**	**Jan. 16**	**Jan. 7**	**Dec. 22**	**16**	**16**	**10.5**	**14.2**
c6, n6, h6	_delay_dd1	No	Jan. 15	Jan. 6	Dec. 22	17.5	17.5	10.5	15.2
c10, n10, h10	**_pp–float_nomins**	**Yes**	**Jan. 22**	**Jan. 15**	**Dec. 23**	**9**	**6**	**8**	**7.7**
c13, n13, h13	_pp–float_400fmin	Yes	22-Jan	Jan. 15	Dec. 23	9	6	8	7.7
c19, n19, h19	**_slider1340–float_400f**	**Yes**	**17-Jan**	**Jan. 12**	**Dec. 23**	**15**	**13**	**8**	**12.0**
c9, n9, h9	_delay_dd2_noppmin	No	Jan. 23	Jan. 13	Dec. 26	6.5	9.5	6	7.3
c7, n7, h7	_delay_dd1_noppmin	No	Jan. 29	Jan. 17	Dec. 27	3	3.5	4.5	3.7
c15, n15, h15	_uro–float_20ppmin	Yes	Jan. 19	Jan. 12	Dec. 27	12.5	13	4.5	10.0
c18, n18, h18	_delay_dd2_slider1340-float	Yes	Jan. 18	Jan. 18	Dec. 28	14	13	3	10.0
c14, n14, h14	_uro–float_400fmin	Yes	Jan. 22	Jan. 15	Dec. 29	9	6	2	5.7
c17, n17, h17	**_slider1340**	**Yes**	**Jan. 23**	**Jan. 17**	**Dec. 31**	**6.5**	**3.5**	**1**	**3.7**

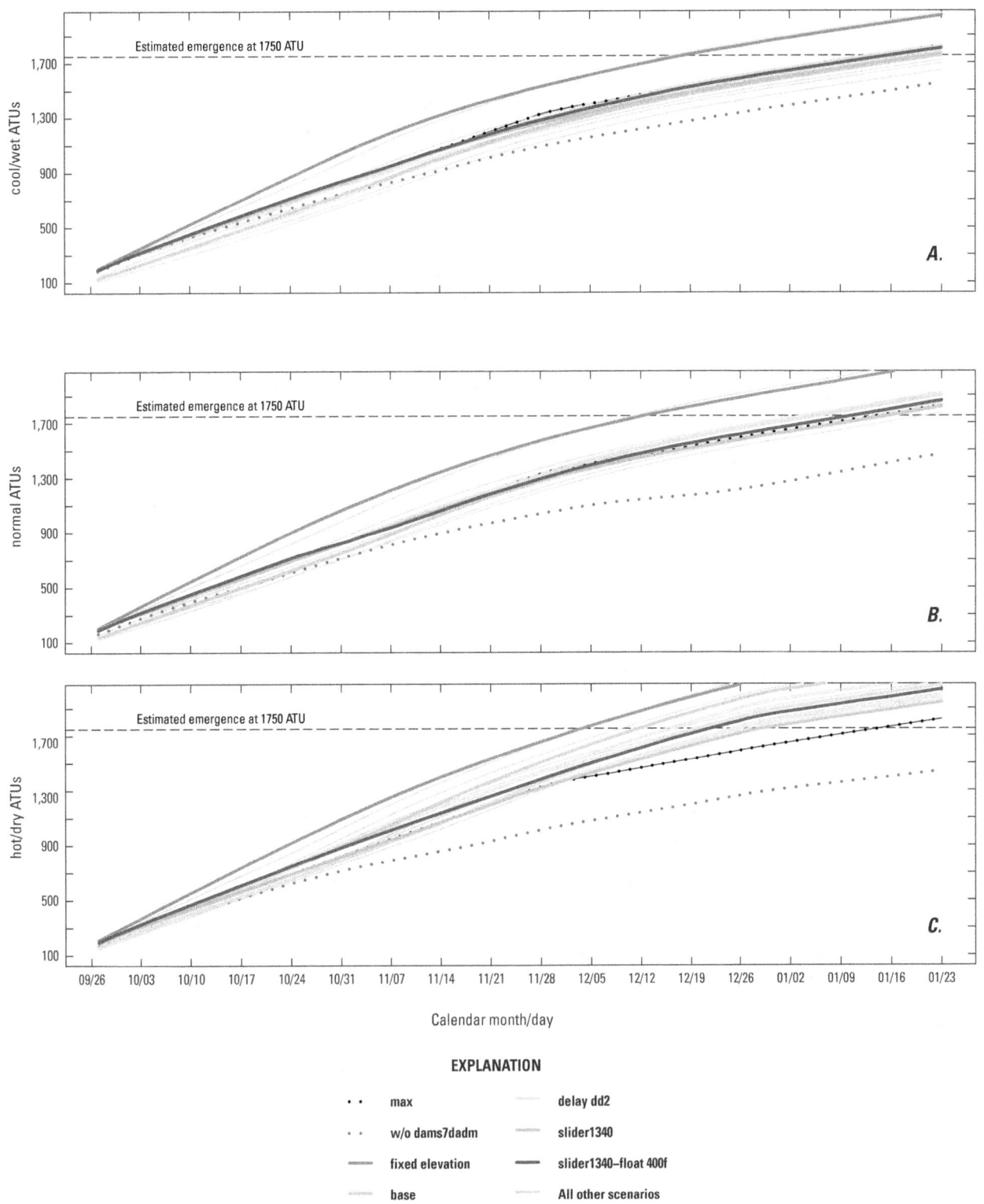

Figure 28. Computed progression of Accumulated Thermal Units (ATUs) for selected model scenarios, showing the different likely rates of egg incubation over the autumn months, North Santiam River, Oregon. Spring Chinook salmon are likely to hatch when the ATU reaches 1,750 degrees Fahrenheit-day. ("_max" is *max* temperature target, "_w/o_dams7dADM" is the 7-day moving average of the daily maximum of the without-dams temperature target.)

Table 9. Calculated emergence day at Mehema (river mile 38.7) from the North Santiam River model were based on the day at which the Accumulated Thermal Units (ATUs) reached 1,750 degrees Fahrenheit-day, North Santiam River, Oregon.

[ATU calculated by the difference of the average daily temperature above 32 degrees Fahrenheit from September 20 through December 31. Dates in January and February were estimated based on model results from each environmental scenario. Scenarios were based on the *max* temperature target]

Scenario identifier	Scenario description	Structural scenario?	Emergence day	Rank
h1	_base	No	Dec. 31	5
h10	_pp-float_nomins	Yes	Jan. 9	4
h17	_slider1340	Yes	Jan. 15	3
h19	_slider1340-float_400f	Yes	Jan. 15	2
h8	_delay_dd2	No	Feb. 3	1

North Santiam River Temperatures

To examine the potential downstream effects of select scenarios at Detroit Dam, outflows from the Detroit Lake model were routed through the downstream Big Cliff Reservoir and North Santiam/Santiam River models. For these simulations, the South Santiam River inflow and all other tributary inflows to the North Santiam River were taken from measured conditions for the environmental scenario of interest. A 7-day moving average of the daily maximum (7dADM) water temperature in each model segment of the North Santiam and Santiam River model was used to display an image of the modeled river temperatures throughout the calendar year of the *hot/dry* environmental scenario.

Base Case

The downstream conditions resulting from current *base* operational rules in place at Detroit Dam under the *hot/dry* environmental forcing conditions show that downstream river temperatures generally increase from the dam release temperatures, except in November and December when river temperatures decrease (fig. 29). Notably, the loss of spillway control at Detroit Dam around August 1 and the change in the temperature target near October 1 are apparent in the figure at RM 59 and can be traced through the river model downstream to the confluence of the North and South Santiam Rivers. The relatively large temperature differences due to operational changes at RM 59 during these times diminish with downstream distance through the river model. The effects of operational changes at Detroit Dam are reduced farther downstream of the confluence of the North and South Santiam Rivers from August through October due to the warming effect of the river system downstream of Big Cliff Dam. In early November, the North Santiam and Santiam Rivers below Big Cliff Dam begin to have a downstream cooling effect.

Delayed Drawdown

By reducing the minimum outflow requirements from Detroit and Big Cliff Dams from September 1 to October 16 as in operational scenario *delay_dd2*, the Detroit Lake water level remained higher than in *base* scenarios (compare figs. 7 and

17). However, this decreased outflow from Detroit and Big Cliff Dams under the *h8* scenario led to downstream warming in the North Santiam River model results that exceeded the *base h1* scenario. Although the outflow temperatures from Big Cliff Dam in *h8* were cooler than those in *h1* during autumn, the reduced flow in the North Santiam River led to temperatures 2–5°C warmer than in the *h1* scenario (fig. 30).

Floating and Fixed-Elevation Gates

The simulation of a hypothetical floating gate combined with the existing power penstock outlets (fixed-elevation gates) at Detroit Dam, with no minimum power-generation requirement (scenario *h10*, table 7), allows the Detroit Lake model to expel warm water during summer beyond the day in which the lake elevation falls below the spillway outlets in the *h1* scenario. During a hot and dry year, the lake level may be lower than *normal* and not allow the use of the spillway late in summer; using a floating outlet circumvents this problem. This scenario has the effect of rationing cool water until later in autumn and cooling the North Santiam River about 1°C compared to scenario *h1* throughout October and the latter half of November (fig. 31). Although that figure shows substantial warming in August in scenario *h10* relative to *h1*, that warming is somewhat artificial because the spillway could no longer be used in the *h1* scenario and cooler-than-desired water was being released in that case.

A Single Sliding-Gate Structure

Additional flexibility and an ability to meet the cool temperature targets in autumn is realized in model scenario *h17* (table 7), in which a sliding-gate outlet ranging from 2 m below the lake surface to the elevation of the upper ROs is simulated at Detroit Dam. Scenario *h17* temperatures between Big Cliff Dam and the South Santiam River confluence were generally 0.5–2.5°C cooler than temperatures for scenario *h1* during October and November (fig. 32). Again, the warming in August shown in figure 32 is somewhat artificial as a result of the loss of the use of the spillway in early August in scenario *h1*; scenario *h17* does a better job of meeting the *max* temperature target at that time. Downstream of the South Santiam River confluence in the Santiam River, scenario *h17* continued to have a cooling effect throughout autumn as water temperatures remained as much as 1°C cooler than the *h1* scenario from mid-October to mid-December.

Floating and Sliding-Gate Structures

Similar to the sliding-gate-only structural scenario *h17*, scenario *h19* specified the use of the same sliding-gate outlet with a lower vertical limit of 1,340 ft, but added a floating outlet with a fixed outflow year-round of 400 ft³/s (table 7). This scenario resulted in autumn release temperatures that were cooler than *h1* (0.5–11.5°C in October and 1.5–3.0°C in November) from RMs 59 to 40 (fig. 33). Downstream of the confluence of the Santiam River near RM 12, the temperature effects from *h19* are similar to *h17*.

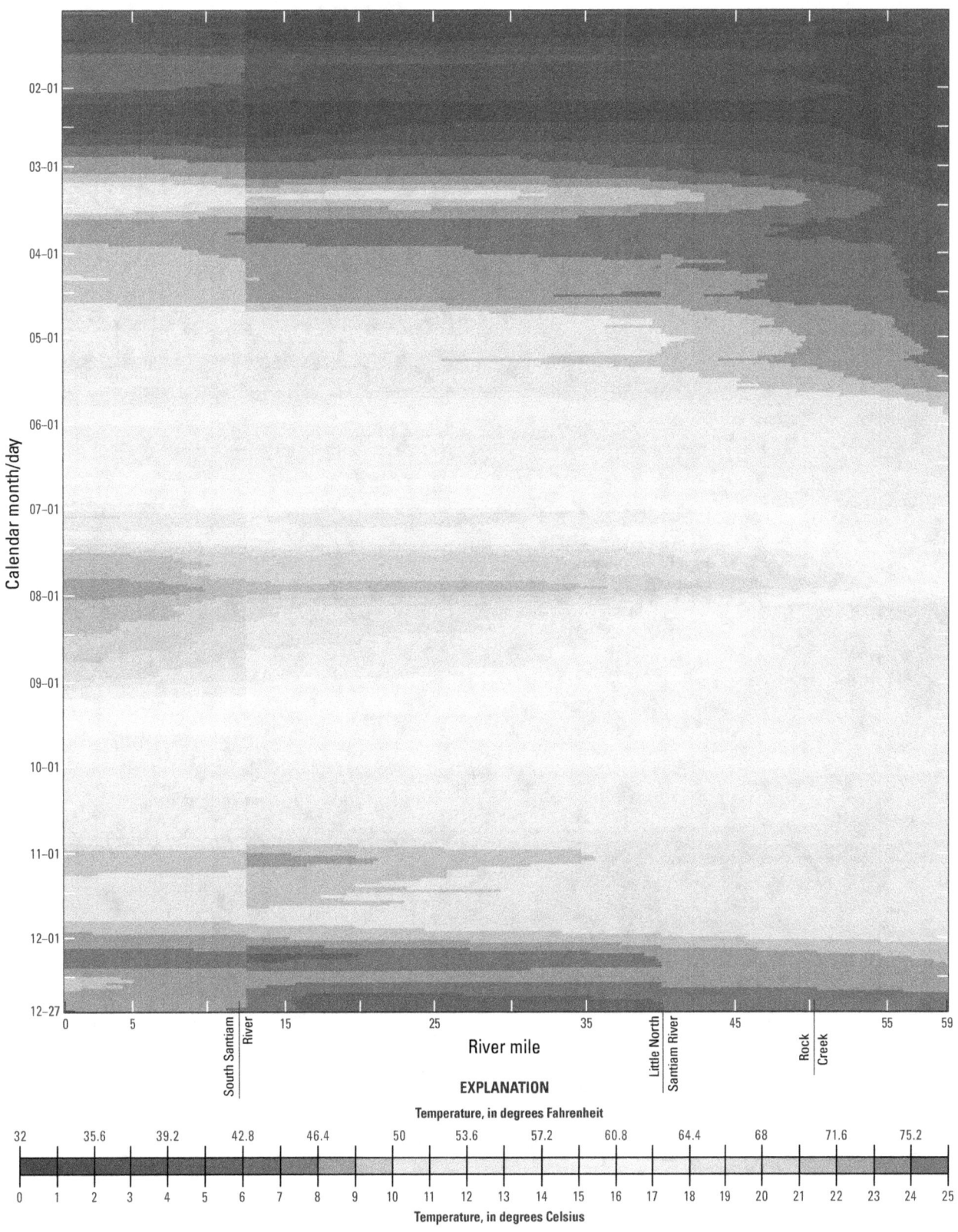

Figure 29. Simulated 7-day moving average of the daily maximum water temperature from the North Santiam and Santiam River model under *hot/dry* environmental conditions and *base* operations.

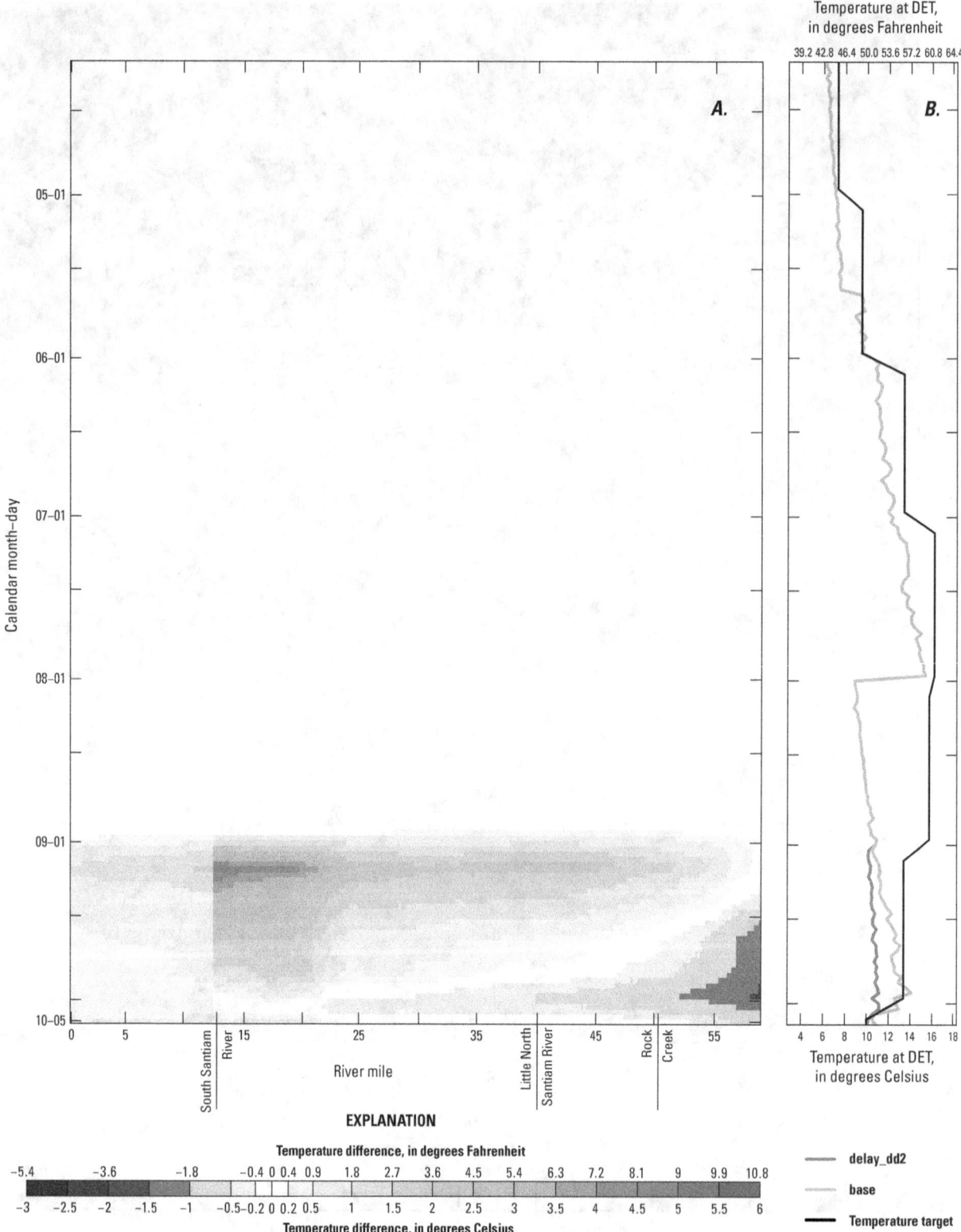

Figure 30. (*A*) Difference between the 7-day moving average of the daily maximum water temperature from the North Santiam and Santiam River model scenarios *h8* (*existing* structures, *delay_dd2* operations, and *hot/dry* conditions) and *h1* (*existing* structures and *base* operations). Positive numbers indicate warming in the *h8* scenario relative to the *h1* scenario; a white color indicates an absolute change of less than 0.2°C. (*B*) Comparison of the max temperature target and simulated outflow temperatures at Detroit Dam from scenarios *h8* (labeled "delay_dd2") and *h1* (labeled "base").

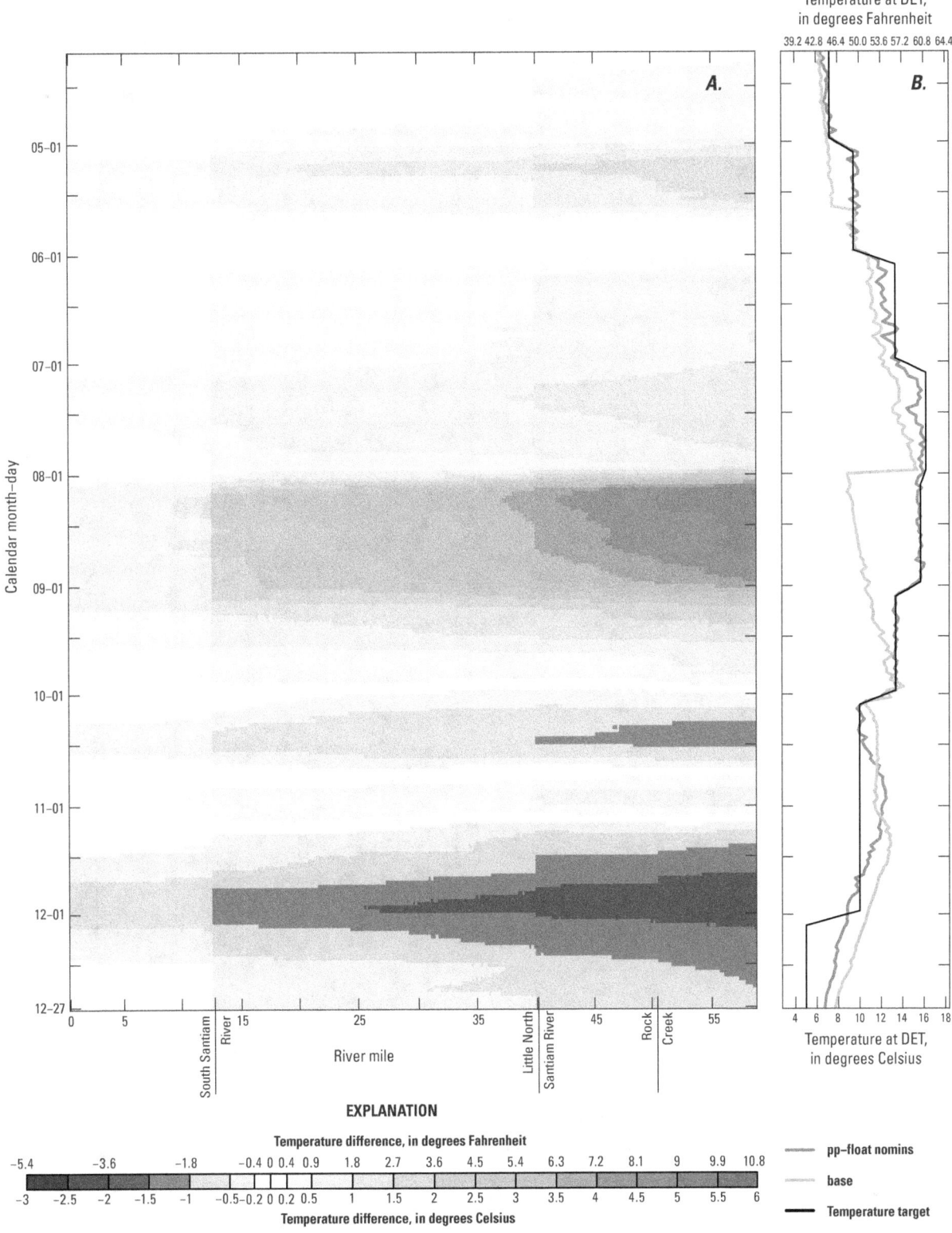

Figure 31. (*A*) Difference between the 7-day moving average of the daily maximum water temperature from the North Santiam and Santiam River model scenarios *h10* (*pp-float* structures, *nomins* operations, and *hot/dry* conditions) and *h1* (*existing* structures and *base* operations). Positive numbers indicate warming in the *h10* scenario relative to the *h1* scenario; a white color indicates an absolute change of less than 0.2°C. (*B*) Comparison of the *max* temperature target and simulated outflow temperatures at Detroit Dam from scenarios *h10* (labeled "pp-float_nomins") and *h1* (labeled "base").

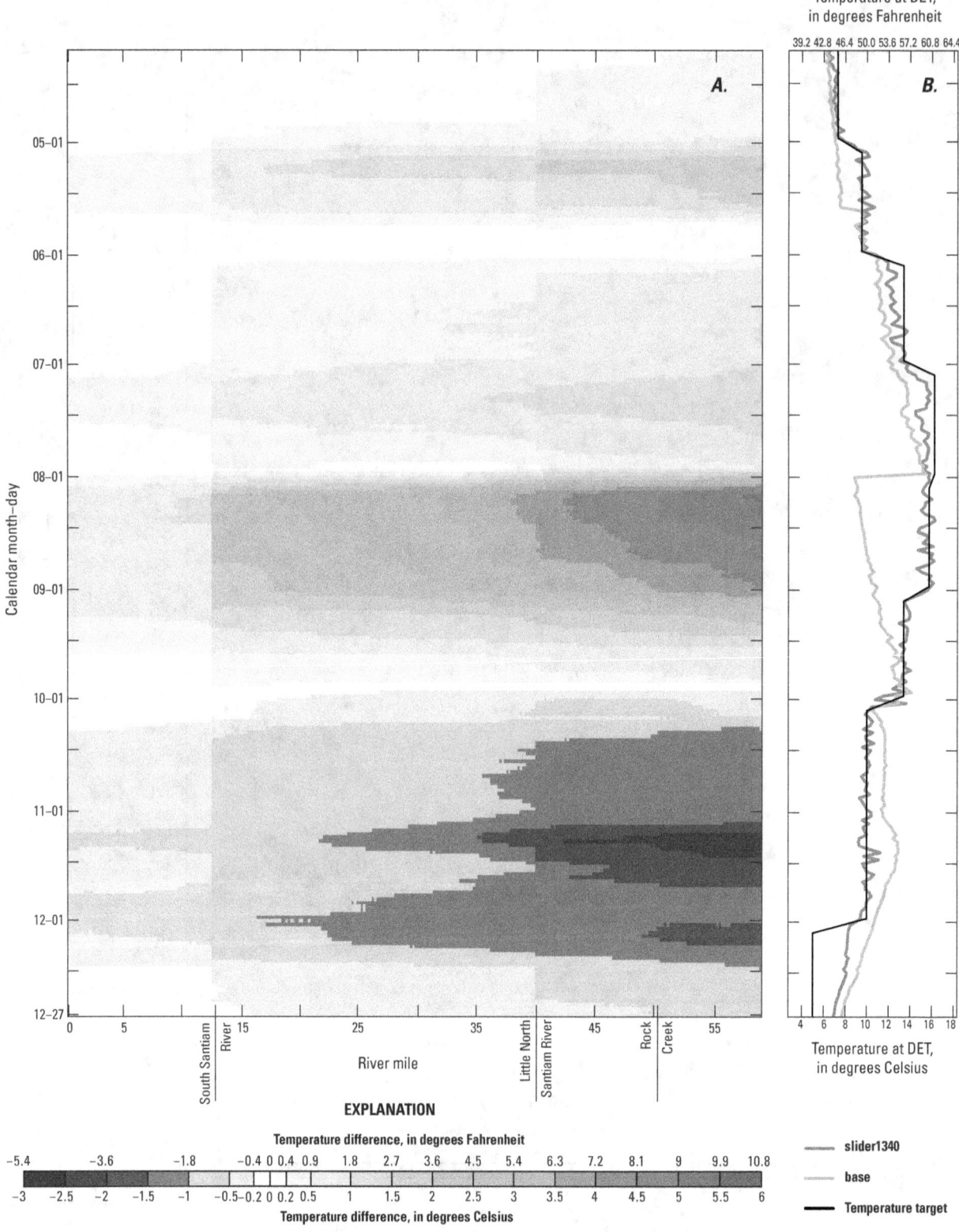

Figure 32.   (*A*) Difference between the 7-day moving average of the daily maximum water temperature from the North Santiam and Santiam River model scenarios *h17* (*slider1340* structures, *base* operations, and *hot/dry* conditions) and *h1* (*existing* structures and *base* operations). Positive numbers indicate warming in the *h17* scenario relative to the *h1* scenario; a white color indicates an absolute change of less than 0.2°C. (*B*) Comparison of the *max* temperature target and simulated outflow temperatures at Detroit Dam from scenarios *h17* (labeled "slider1340") and *h1* (labeled "base").

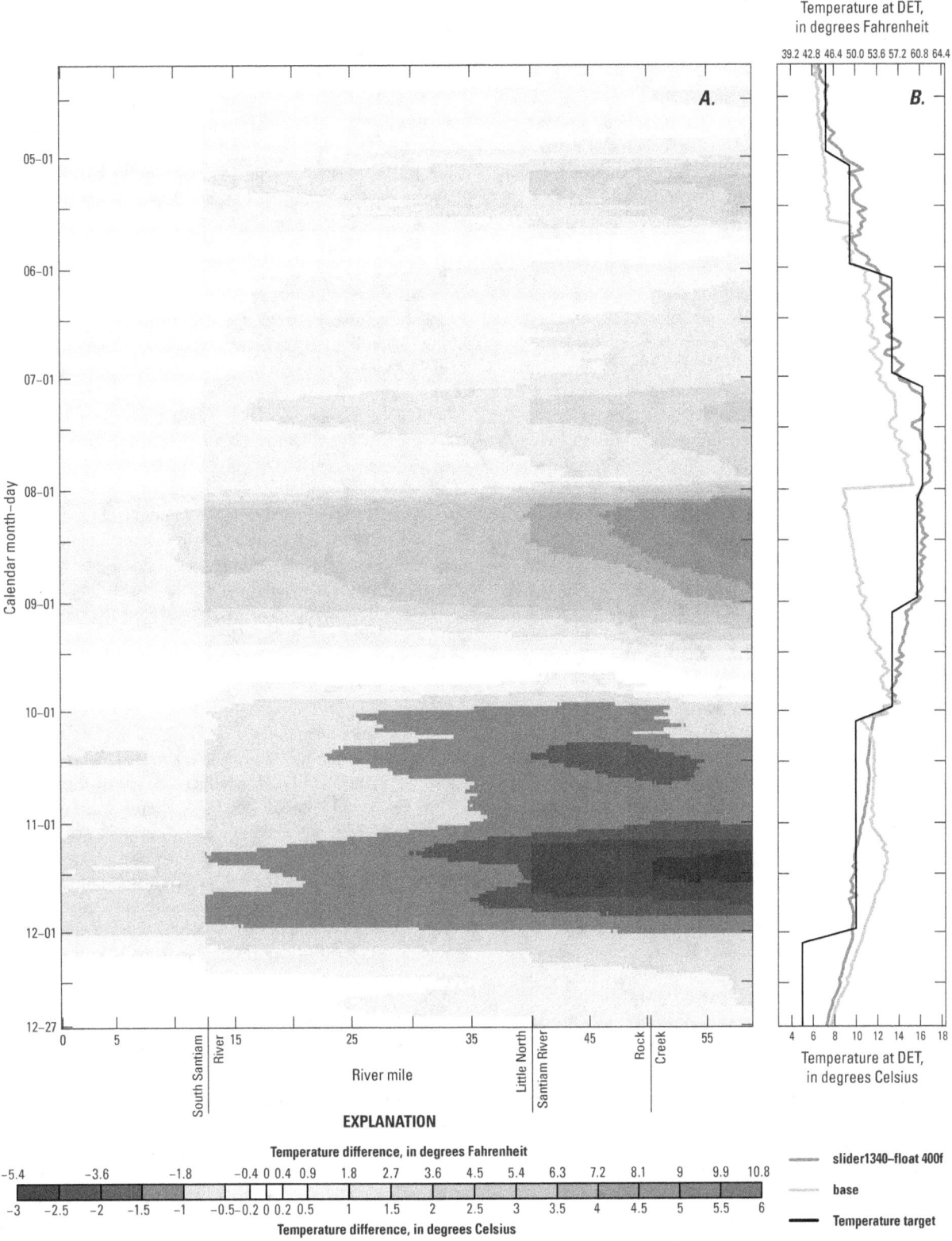

Figure 33.   (*A*) Difference between the 7-day moving average of the daily maximum water temperature from the North Santiam and Santiam River model scenarios *h19* (*slider1340-float* structures, *400f* operations, and *hot/dry* conditions) and *h1* (*existing* structures and *base* operations). Positive numbers indicate warming in the *h19* scenario relative to the *h1* scenario; a white color indicates an absolute change of less than 0.2°C. (*B*) Comparison of the *max* temperature target and simulated outflow temperatures at Detroit Dam from scenarios *h19* (labeled "slider1340-float _400f") and *h1* (labeled "base").

Implications for Dam Operation and Planning

The Detroit Dam model results from this study show the range of release temperatures that might occur under varying hydrologic and meteorological conditions as well as under several operational and structural scenarios. A common theme among all model results is that spring and summer dam operations have an effect on operational flexibility and the extent to which release temperatures can be controlled later in autumn. Model results indicate that as early in the year as April, solar radiation heats the surface of the lake and thermal stratification begins. Because most of the lake vertical profile is still relatively cool at that time, the ability to meet downstream temperature targets during spring is dependent on the ability to access and release warmer water near the lake surface. This can be difficult when the lake surface is either well below or well above the spillway crest in spring and early summer. As the surface of the lake becomes warmer throughout summer and the thermocline moves deeper in the lake, access to cool water below the thermocline begins to decrease from about June until about mid-November, at which point the lake has been drawn down to make room for potential flood storage and typically is isothermal. In general, the release of warm surface water from the lake during summer allows the cooler water deeper in the lake to be reserved until autumn when that cold water is needed most to meet downstream temperature targets.

The ability to mix and release (warm) lake surface water with (cold) deeper water throughout the year often is the limiting factor in controlling release temperatures from Detroit Lake or other deep reservoirs with similar outlet configurations. The existing outlets at Detroit Dam do not allow near-surface waters to be released during times when the lake elevation is below the spillway crest (spring and autumn). During years in which the reservoir may be late to fill or not fill at all (as in *hot/dry* and *base* model scenarios), the spillway may only be a viable release point for a limited time in summer. Immediately after the lake is drawn down below the spillway crest elevation, dam operations with existing outlets must release cool water from below the thermocline using either the power penstocks or the upper ROs. Later in the year, the cool water supply below the thermocline can become exhausted at the elevation of the available outlets, and an uncontrollable rise in release temperatures results from about October through November. Thus, the existing structures allow the managers and operators of Detroit Dam to blend releases from multiple outlets for only part of the year, with less flexibility in drier years.

Power production requirements limit the use of existing structures at Detroit Dam to expel warmer water in summer (June–September) and cooler water in autumn (October–November). Operational scenarios with no minimum outflow requirements to the power penstocks (*noppmin*) led to outflow temperatures from Detroit Dam that were closer to meeting downstream temperature targets than operations with dedicated minimum flows for power production.

Model simulations indicate that by delaying the drawdown of Detroit Lake in autumn, better control over release temperatures is possible immediately downstream of Big Cliff Dam. Delaying the drawdown of the lake for better downstream temperature management must be balanced against the need to make room in the reservoir to manage storm-related November inflows that might lead to floods. The temperature benefits of delaying lake drawdown result mainly from an extended use of the Detroit Dam spillway until as late as November 1 (see *delay_dd1* and *delay_dd2* scenarios in figs. 14 and 17). By delaying the date at which drawdown begins, warm epilimnetic water can continue to be released in conjunction with cool water from the hypolimnion, thereby rationing the deeper cool-water supply throughout autumn. As a result of this sustained use of the spillway under these operational scenarios (figs. 15, 16, 18, and 19), the abrupt change in release temperature caused by the loss of spillway usage is not as apparent as with *base* operational scenarios (fig. 8). Such abrupt changes in release temperatures may or may not be a consideration for downstream salmon habitat during late summer and autumn. Farther downstream of Big Cliff Dam, however, the North Santiam River model results show that decreased releases during September 1–October 15 necessary to keep lake levels high in scenario *h8* (see *delay_dd2* operations in table 4) cause substantial downstream warming (2–5°C).

Aside from operationally delaying the drawdown of Detroit Lake, a number of simulated structural scenarios showed that the addition of hypothetical floating outlets at Detroit Dam could provide access to warm surface water to be released spring through autumn, allowing better management of release temperatures throughout the season. Adding a floating outlet generally leads to greater control of the outflow temperature compared with existing outlets at Detroit Dam, even under *hot/dry* conditions. Combining the upper ROs with a floating outlet (*uro-float*) resulted in greater temperature control in autumn than the combination of the power penstocks and a floating outlet (*pp-float*) (compare figs. 23 and 24 or fig. 22 with 25). As the elevation of the lower outlet was decreased (going from the power penstock elevation to the upper RO elevation), the amount of outflow temperature control at Detroit Dam increased. As decreased minimum flow requirements were placed on the lower, fixed-elevation outlets in these scenarios (that is, an increase in the allowable "percent spill"), the resulting outflows generally were cooler in autumn. Likewise, warmer outflows during June and July were possible under these scenarios and may have contributed to the relatively large supply of accessible cool water in the lake later in autumn.

When a hypothetical sliding outlet was used alone (*slider1340*), outflow temperatures roughly met the *max* temperature target at Detroit Dam (fig. 20), yet this scenario resulted in more day-to-day temperature variation than equivalent scenarios incorporating both a floating and lower (fixed-elevation or sliding-gate) outlet. This illustrates the value provided by having two outlets to access warm and cold water separately throughout the year. As the thermocline moves up and down in the water column during the day due to seiching, a more variable release temperature results from a single sliding-gate outlet (*slider1340)* than from a blended combination of one floating outlet withdrawing warmer surface water and one fixed-elevation outlet withdrawing cooler water (*slider1340-float)*.

The estimated emergence date of spring Chinook salmon was tabulated as a way of comparing the relative success of the model scenarios in this study. Success, as measured in this study, is a delay in the estimated spring Chinook emergence date, as early emergence can be problematic (U.S. Army Corps of Engineers, 2012). The comparison showed that many structural scenarios and scenarios in which no minimum flow was directed to the power penstocks generally led to later emergence dates. Under *hot/dry* environmental forcing conditions, structural scenarios generally exhibited later emergence dates than scenarios incorporating only operational changes to Detroit Dam, perhaps in large part because of the limited use of the spillway under *hot/dry* conditions in non-structural scenarios. Downstream in the North Santiam River, estimated emergence dates also were influenced by cool inflows from large tributaries such as the Little North Santiam River, which delayed the emergence date significantly. The emergence date is not the only factor involved in assessing the biological success of an operational or structural scenario, however, as with other streamflow and habitat considerations, may be important.

Results from the Detroit Lake model show that the ability to control release temperatures and meet downstream temperature targets throughout the year can be more closely attained at the site of the dam by delaying drawdown of the lake in autumn, decreasing the minimum power-generation requirement during summer/autumn, and (or) installing a well-conceived combination of floating and (or) sliding-gate outlets.

Results from the North Santiam and Santiam River model downstream of Big Cliff Dam show that release temperatures from Detroit and Big Cliff Dams have an important and measurable effect in the North Santiam River, especially in autumn as days become shorter and solar radiation imposes less heating to the river. The river modeling illustrated the importance of both flow rate *and* water temperature

downstream of Big Cliff Dam during autumn to benefit spring Chinook salmon spawning. The temperature effects of altered releases at Detroit Dam tend to diminish with downstream distance, but the effects are large enough to be measurable throughout the North Santiam and Santiam River systems. The temperatures and seasonal temperature pattern downstream of Detroit Dam in the North Santiam River system can be managed and controlled through a variety of changes in dam operations or outlet options at the upstream dams.

Acknowledgments

This study was made possible with primary funding from the U.S. Army Corps of Engineers, Portland District. Kathryn Tackley, James Britton, Mary Karen Scullion, Ian Chane, and Cindy Bowline (USACE) provided data, helpful discussions, and strategic details and direction on potential options for dam operations and structural possibilities. Steven Sobieszczyk (USGS) assisted with geographic information system (GIS) work.

References Cited

Buccola, N.L., and Rounds, S.A., 2011, Simulating potential structural and operational changes for Detroit Dam on the North Santiam River, Oregon—Interim results: U.S. Geological Survey Open-File Report 2011–1268, 32 p. (Also available at http://pubs.usgs.gov/of/2011/1268/.)

Caissie, D., 2006, The thermal regime of rivers—A review: Freshwater Biology, v. 51, p. 1389–1406.

Cole, T.M., and Wells, S.A., 2002, CE-QUAL-W2—A two-dimensional, laterally averaged, hydrodynamic and water-quality model, version 3.1: U.S. Army Corps of Engineers, Instruction Report EL-02-1 [variously paged].

Cole, T.M., and Wells, S.A., 2008, CE-QUAL-W2—A two-dimensional, laterally averaged, hydrodynamic and water-quality model, version 3.6: U.S. Army Corps of Engineers, Instruction Report EL-08-1 [variously paged].

Moore, A.M., 1964, Compilation of water-temperature data for Oregon streams: U.S. Geological Survey Open-File Report 64-115, 134 p., 1 pl. (Also available at http://pubs.er.usgs.gov/usgspubs/ofr/ofr64115.)

Moore, A.M., 1967, Correlation and analysis of water-temperature data for Oregon streams: U.S. Geological Survey Water-Supply Paper 1819-K, 53 p., 1 pl. (Also available at http://pubs.er.usgs.gov/usgspubs/wsp/wsp1819K.)

Nash, J.E., and Sutcliffe, J.V., 1970, River flow forecasting through conceptual models part I—A discussion of principles: Journal of Hydrology, v. 10, no. 3, p. 282–290.

National Marine Fisheries Service, 2008, Willamette Basin Biological Opinion—Endangered Species Act Section 7(a)(2) Consultation: National Oceanic and Atmospheric Administration Fisheries Log Number F/NWR/2000/02117 [variously paged], accessed October 20, 2009, at http://www.nwr.noaa.gov/Salmon-Hydropower/Willamette-Basin/Willamette-BO.cfm.

Oregon Department of Environmental Quality, 2006, Willamette Basin Total Maximum Daily Load program documents: Oregon Department of Environmental Quality, accessed March 27, 2007, at http://www.deq.state.or.us/wq/TMDLs/willamette.htm.

Oregon Department of Environmental Quality, 2009, Water quality standards—Beneficial uses, policies, and criteria for Oregon—Temperature: Oregon Administrative Rule 340-041-0028: Oregon Department of Environmental Quality, accessed November 13, 2009, at http://arcweb.sos.state.or.us/pages/rules/oars_300/oar_340/340_tofc.html#041.

Rounds, S.A., 2010, Thermal effects of dams in the Willamette River Basin, Oregon: U.S. Geological Survey Scientific Investigations Report 2010–5153, 64 p. (Also available at http://pubs.usgs.gov/sir/2010/5153/.)

Rounds, S.A., and Sullivan, A.B., 2006, Development and use of new routines in CE-QUAL-W2 to blend water from multiple reservoir outlets to meet downstream temperature targets, in Proceedings of the Third Federal Interagency Hydrologic Modeling Conference, April 2–6, 2006, Reno, Nev.: Subcommittee on Hydrology of the Interagency Advisory Committee on Water Information, ISBN 0-9779007-0-3, 8 p., accessed September 11, 2012, at http://or.water.usgs.gov/tualatin/fihmc3_w2_modifications.pdf.

Sullivan, A.B., and Rounds, S.A., 2004, Modeling streamflow and water temperature in the North Santiam and Santiam Rivers, Oregon: U.S. Geological Survey Scientific Investigations Report 2004–5001, 35 p. (Also available at http://pubs.usgs.gov/sir/2004/5001/.)

Sullivan, A.B., and Rounds, S.A., 2006, Modeling water-quality effects of structural and operational changes to Scoggins Dam and Henry Hagg Lake, Oregon: U.S. Geological Survey Scientific Investigations Report 2006–5060, 36 p. (Also available at http://pubs.er.usgs.gov/publication/sir20065060.)

Sullivan, A.B., Rounds, S.A., Sobieszczyk, S., and Bragg, H.M., 2007, Modeling hydrodynamics, water temperature, and suspended sediment in Detroit Lake, Oregon: U.S. Geological Survey Scientific Investigations Report 2007–5008, 40 p. (Also available at http://pubs.er.usgs.gov/publication/sir20075008.)

U.S. Army Corps of Engineers, 2012, Willamette Basin Annual Water Quality Report for 2011: U.S. Army Corps of Engineers Final Report, April 2012, 123 p.

Appendix A.   Big Cliff Reservoir Model Development—Construction and Calibration

Abstract

A hydrodynamic and water temperature model was developed for Big Cliff Reservoir on the North Santiam River in western Oregon for calendar years 2002 and 2003. This model allows the connection of an existing model of Detroit Lake upstream to an existing model of the North Santiam River downstream. The Big Cliff Reservoir model was able to reproduce the daily as well as hourly fluctuations in water surface elevation well. Initial runs showed that the magnitude and seasonal patterns in modeled water temperature released from Big Cliff Dam matched measured temperature just downstream in the North Santiam River generally well; however, model temperatures were 2 to 3°C too warm in late October to early November. Sensitivity testing and other investigations into this issue led to modifications in the setup of the modeled Detroit Lake model releases, which formed the upstream boundary of the Big Cliff Reservoir model. These changes led to somewhat higher water temperature errors within the Detroit Lake model, but improved the measured-to-modeled fit for the Big Cliff release in late October to early November in both 2002 and 2003.

Introduction

Big Cliff Reservoir and Big Cliff Dam are part of the USACE water management system in the Willamette River basin in northwest Oregon (fig. 1). Big Cliff Dam was constructed in 1953 along with the larger Detroit Dam, about 2.8 mi upstream. At a full-pool water surface elevation of 1,206 ft, Big Cliff Reservoir stores 6,450 acre-ft of water. Big Cliff Dam releases water through a power generating facility or through radial spillway gates to the North Santiam River.

A primary purpose of Big Cliff Dam is to regulate the fluctuating power-generating water releases from Detroit Dam so that relatively smooth flows are released to the North Santiam River. In the years modeled, the Big Cliff Reservoir water surface elevation often fluctuated on a daily or hourly basis as much as 24 ft due to the hydropower peaking releases from Detroit Dam. Other purposes of Big Cliff Reservoir and Dam include flood damage protection, power generation, water quality improvement, fish and wildlife habitat, and recreation.

Purpose and Scope

The purpose of this work was to develop a model of Big Cliff Reservoir that could (1) simulate stage, flow, velocity, and water temperature, (2) provide information on processes that control water temperature in this reach, and (3) act as the connecting model between the existing Detroit Lake CE-QUAL-W2 model and the existing North Santiam and Santiam River CE-QUAL-W2 model so that model scenarios for the entire system could be run and analyzed. Separate Big Cliff Reservoir models were developed for calendar years 2002 and 2003 and were calibrated for flow and water temperature.

Methods and Data

The Big Cliff Reservoir model was constructed using version 3.12 of CE-QUAL-W2, a hydrodynamic and water quality model from the USACE (Cole and Wells, 2002). CE-QUAL-W2 is two-dimensional, simulating vertical and longitudinal variation from upstream to downstream; it is laterally averaged across the channel. For a long, narrow, pooled reach such as Big Cliff Reservoir, a two-dimensional laterally averaged model is a good choice. CE-QUAL-W2 can simulate streamflow, water velocity, water temperature, and a number of water quality constituents, including total dissolved solids, nutrients, algae, oxygen, and suspended sediment. CE-QUAL-W2 also was used to build the upstream Detroit Lake model (Sullivan and others, 2007) and the downstream North Santiam River model (Sullivan and Rounds, 2004) as well as models of other rivers and reservoirs in the Willamette River basin.

The CE-QUAL-W2 model code was modified by USGS project personnel to (1) fix coding errors, (2) add new model flux outputs, (3) add a new subroutine to automatically blend outflows from multiple reservoir outlets to match a user-supplied downstream temperature target, and (4) update the selective withdrawal algorithms. The blending routines were documented previously by Sullivan and Rounds (2006); further updates are documented in appendix C.

The Big Cliff Reservoir model was constructed in several steps. Initially, a model grid was built. Then, model input data were collected, processed, and formatted to provide flow, water temperature, meteorological, and shade boundary conditions. Finally, the model was calibrated by comparing model output to measured water surface elevation and water temperature data.

Model Grid

A CE-QUAL-W2 model grid is composed of model segments that connect together in the direction of flow. Each individual segment has layers with defined height that increase in width from the channel bottom upwards, resembling a cross section in shape. Only limited bathymetric data were available to construct the Big Cliff model grid. As a first step, a pre-dam topographic map from USACE of the Big Cliff reach with only few contour lines was digitized into a geographic information system (GIS). Model segment boundaries were designated in this GIS coverage. Then, 10 equally spaced cross sections were sampled within each segment using GIS techniques; the ten subsections were averaged to obtain a representative cross section shape for each model segment. Layer widths were adjusted until the volume-elevation curve matched the volume-elevation curve from the USACE Big Cliff Reservoir storage table (USACE table dated November 15, 2002) (fig. A1). After checking that none of the layer widths were less than 5 m, as recommended by the CE-QUAL-W2 development team (Cole and Wells, 2008), the segment cross sections were formatted into a CE-QUAL-W2 bathymetry input file. The Big Cliff CE-QUAL-W2 model grid consisted of 15 model segments. Segment length ranged from 281.2 to 307.0 m, with an average length of 297.3 m. Layer height throughout the bathymetry grid was 1.0 m.

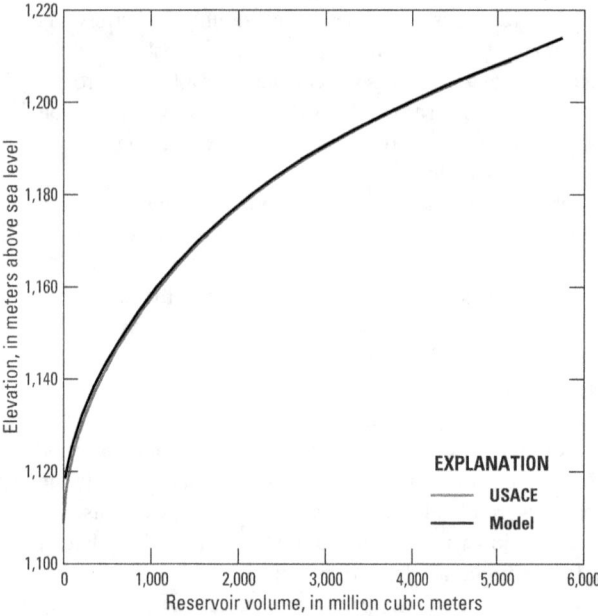

Figure A1. Volume-elevation curves for Big Cliff Reservoir, Oregon, from the U.S. Army Corps of Engineers (USACE) and as represented by the model grid (Model).

Model Data

Because the Detroit Lake and Big Cliff Reservoir models are adjacent, some CE-QUAL-W2 inputs from the Detroit Lake model could be used for the Big Cliff Reservoir model. Shared inputs included the meteorological conditions input file, the precipitation input file, the precipitation temperature input file, and many control file parameters. Data for other input files and calibration had to be obtained specifically for the Big Cliff model. Although the Detroit Lake model simulated total dissolved solids and suspended sediment, those constituents were not included in the Big Cliff model because the North Santiam River model does not simulate those constituents.

Hydrologic Data

The main inflow to the Big Cliff Reservoir model was the outflow from the Detroit Lake model. These releases often fluctuated greatly on an hourly basis, from zero to hundreds or thousands of cubic feet per second. Flows from Detroit Dam were released in this manner to respond to hydropower demands. During high flow events, flows were released from Detroit Dam in a more continuous fashion.

Other inflows to Big Cliff Reservoir included tributary flows from Sardine Creek and Lawhead Creek. Because neither of those inflows were gaged, the inflows were estimated by multiplying the ratio of each creek's watershed (drainage) area to the watershed area of Blowout Creek by the gaged flow of Blowout Creek. Blowout Creek is a gaged tributary on the south side of Detroit Lake that has a long record of data. Sardine Creek drains 5.5 mi^2 and Lawhead Creek drains 4.6 mi^2 of watershed area.

Water releases from Big Cliff Dam were routed through the power penstocks or over the spillway. Data on hourly flow through these two outlets were obtained from USACE. The power penstocks' intake centerline elevation is 1,140 ft (347.5 m), and the spillway crest is located at 1,161.5 ft (354.0 m). These water release elevations were set in the model control file.

The measured water surface elevation of Big Cliff Reservoir was used to close the water balance during model calibration and set the inflows from other ungaged tributaries and groundwater as a distributed tributary input to the model. Hourly values of the Big Cliff Reservoir forebay water surface elevation were obtained from USACE. In 2002–03, the water surface elevation in Big Cliff Reservoir fluctuated between 1,181.0 and 1,205.9 ft.

Water Temperature Data

The water temperature for the inflow from Detroit Lake came from Detroit Lake model output. Water temperatures of Sardine Creek and Lawhead Creek were not measured, but they were estimated to be similar to that of French Creek, a tributary in the Detroit Lake drainage.

Water temperature data with which to compare modeled water temperature during calibration was limited. For instance, during the years modeled, no in-reservoir temperature profiles had been collected. Measured water temperature data were available for the North Santiam River at Niagara, about 0.7 mi downstream of Big Cliff Dam. In addition, some intermittent measured temperature data were available from the base of Detroit Dam from the Oregon Department of Environmental Quality LASAR database.

Some water temperature data in Big Cliff Reservoir for more recent years (parts of 2008 and 2009) were provided by USACE. Although these data were not used directly to calibrate the 2002 and 2003 Big Cliff Reservoir models, the data were useful for helping to understand the general trend in water temperature as water moved from Detroit Dam through Big Cliff Reservoir and farther downstream.

Shade

Because Big Cliff Reservoir is located in a canyon, the effect of topographic shading was important and included in the model simulations. Eighteen topographic inclination angles, every 20 degrees, from the water surface of each model segment to the nearby ridgetops, were calculated using GIS techniques. These angles then were formatted into a CE-QUAL-W2 shade file to describe the shading provided by topographic features. Shading provided by any riparian vegetation was assumed to be negligible.

Model Development/Calibration

During the process of model calibration, measured data are compared to model outputs. Parameters and other factors can be modified within reasonable bounds to optimize the comparison between model outputs and measured data for this specific reach. For the Big Cliff Reservoir model calibration, the water balance was completed first. Then, the model was calibrated for water temperature.

Water Balance

Results from initial model runs that included inflows, outflows, precipitation, and evaporation showed differences between modeled and measured water surface elevations in Big Cliff Reservoir. This indicated that some additional inflows or outflows were needed to close the water balance. Typically, missing flows in a CE-QUAL-W2 model occur due to the presence of small ungaged tributaries, overland flows, groundwater sources or sinks, or error in the measurement of the included inflows and outflows.

For the Big Cliff model, a distributed tributary was used to describe and include these missing flows; this is a common way to close the water balance in CE-QUAL-W2 (Cole and Wells, 2002). In brief, the distributed tributary flow was calculated by subtracting the sum of inflows from the sum of outflows on an hourly basis, applying a moving daily average to that time series, running the model with that distributed tributary file, and making minor adjustments to the distributed tributary inputs until the measured and modeled water surface elevations matched reasonably well. The flow associated with the distributed tributary was small relative to total inflows and outflows, accounting for only 1 and 4 percent of total inflows and 1 and 0 percent of total outflows in 2002 and 2003, respectively. The flows that make up the distributed tributary flows are likely sourced mostly from surface water because the flow imbalance was greatest during storm events. The final modeled water surface elevations were in good agreement with the measured values for both 2002 and 2003 (fig. A2). The water-surface elevations in that figure show a large amount of daily variation, which is typical of how Big Cliff Reservoir is used to moderate (reregulate) the greatly varying releases from Detroit Dam.

Water Temperature

Initial Testing of Big Cliff Reservoir Model

After the water balance was complete, the modeled water temperature of the Big Cliff Dam release was compared to measured water temperature 0.7 mi downstream at the USGS gaging station at Niagara on the North Santiam River. Travel time is short between these locations and although the water temperatures would not be expected to match exactly, they were likely to be close. In this first comparison, the seasonal pattern of water temperature from the modeled Big Cliff release matched the seasonal pattern in the measured data at Niagara for most of the year. However, the annual maximum modeled temperature for the period from late October to early November was as much as 2 or 3°C warmer than the measured temperature.

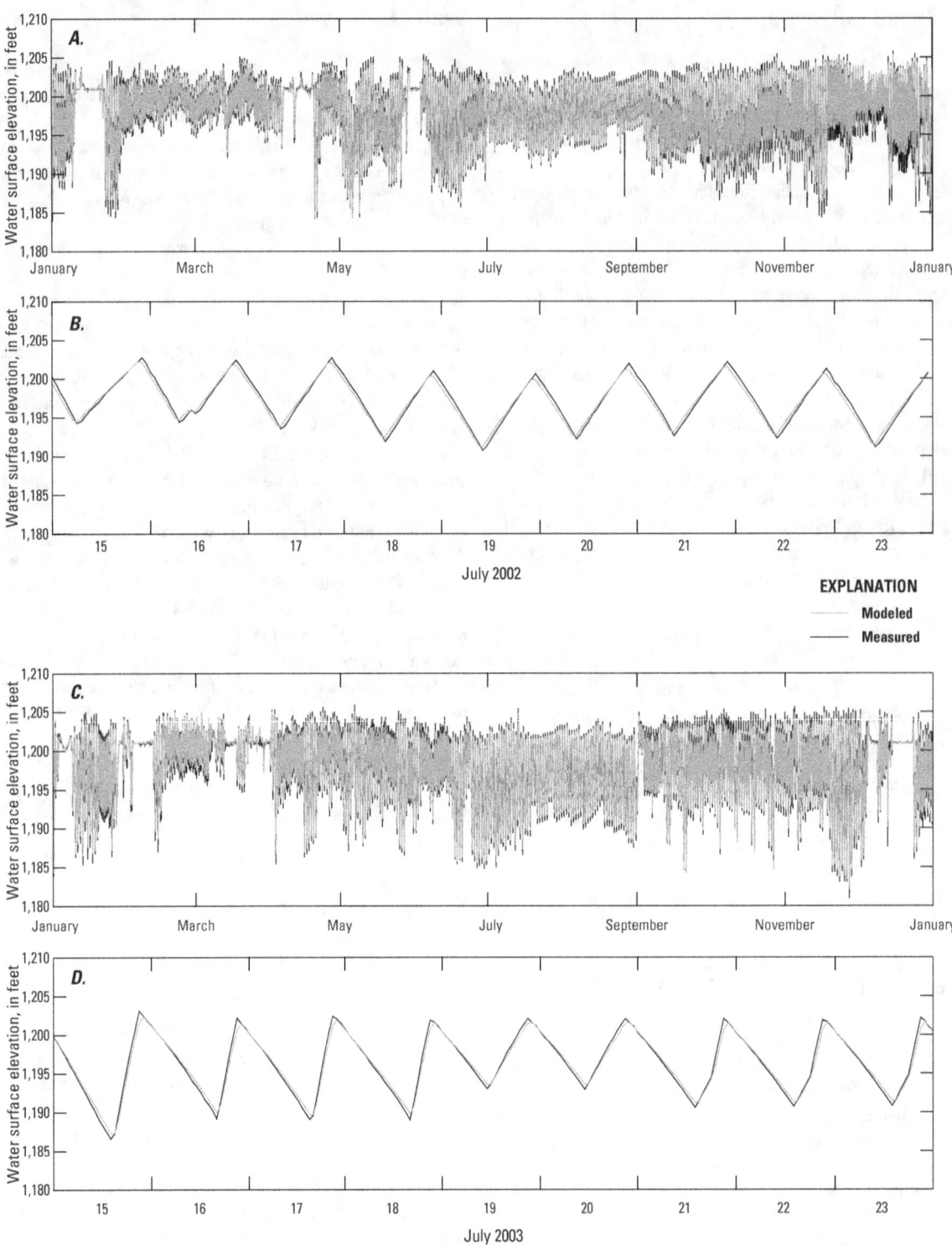

Figure A2. Modeled and measured Big Cliff Reservoir water surface elevations for the entire calendar years of 2002 and 2003. A closer look at 9 days in July shows the nature of the daily variation in water surface elevation.

To test whether model calibration factors within the Big Cliff model could be adjusted to provide a better water temperature match in late October and early November, a series of sensitivity tests were run. The sensitivity tests modified one factor at a time and examined the effect on Big Cliff outflow water temperatures. Factors examined in this analysis included friction factors, the coefficient of bottom heat exchange, the surface heat exchange calculation method, the vertical turbulence closure algorithm, and the elevation of the outflows at Big Cliff Dam. Version 3.6 of CE-QUAL-W2 (Cole and Wells, 2008) also was tested. None of these tests could explain the late October to early November temperature difference and most produced less than a 0.3°C change in the Big Cliff Dam outflow water temperature. The insensitivity of Big Cliff release temperatures to Big Cliff model parameters was likely due to the short model reach and brief residence time of water within Big Cliff Reservoir.

Detroit Lake Model Tests and Adjustments

Because Big Cliff Reservoir model parameters could not explain the 2–3°C discrepancy in late October and early November, the next step was to look farther upstream at the Detroit Lake model and its outflow water temperature. Testing of the Detroit Lake model first took the form of sensitivity testing for parameters that affected temperature both within Detroit Lake and for the Detroit Lake outflow. Through this testing, it was determined that the modeled in-lake water temperature and associated water temperature parameters were constrained by calibration to the in-lake data; therefore, the main variable that could be adjusted was the setup of the Detroit Dam outlet structures and the interaction of the withdrawal outlets with the CE-QUAL-W2 selective withdrawal algorithm.

To address this, several updates and adjustments were made to the CE-QUAL-W2 code for the Detroit Lake model. First, the DOWNSTREAM_WITHDRAWAL and LATERAL_WITHDRAWAL subroutines in the USGS-modified version 3.12 model were modified to make the velocity profile equations similar to those in CE-QUAL-W2 version 3.6. Secondly, the LATERAL_WITHDRAWAL subroutine was modified to allow both point and line withdrawals, using equations from the DOWNSTREAM_WITHDRAWAL subroutine. A point withdrawal is an outlet structure that is narrow in relation to the dam width, whereas a line withdrawal is wide in relation to dam width ($>1/10$). In the previous version of the Detroit Lake model code, the default was to specify point withdrawal outlets at Detroit Dam, which has no associated width specification. A line withdrawal, on the other hand, requires an associated outlet width to be specified, and varying the width of the outlet line affects which lake depths (or model layers) from which the resulting outflow are drawn. If the lake is well-mixed with similar temperatures from surface to bottom, the outlet line width has little effect on outflow water temperature; however, if the lake is stratified with variable water temperature with depth, then this parameter does affect the outflow water temperature.

More specifically, the equations for point and line withdrawals are (Cole and Wells, 2008):

$$\text{point: } d = \left(\frac{c_{bi} Q}{N}\right)^{0.3333}$$

$$\text{line: } d = \left(\frac{c_{bi} 2q}{N}\right)^{0.5}$$

where

d is withdrawal zone half height, m;

Q is total outflow, in m^3/s;

N is internal buoyance frequency, Hz;

q is outflow per unit width, m^2/s; and

c_{bi} is boundary interference coefficient.

As the outlet line width increases, the model withdraws more of its releases from model layers (or reservoir depths) close to the elevation of the outlet. As the line width decreases, the model withdraws water from a greater range of depths. Similarly, greater release rates tend to draw water from more model layers, whereas small releases tend to be from layers near the outlet elevation. Changing the line width then, changes release water temperatures during stratified conditions in Detroit Lake. For the Detroit Lake model, the line width was used as a calibration parameter to better match the late October to early November water temperature downstream at Niagara. The final structure widths used for the Detroit Lake model were 6.8 m for the power penstocks and 4.0 m for the upper ROs; the spillway was not used in 2002–03. These are calibration parameters, and the selective withdrawal algorithms in the model are not perfect representations of mixing near the dam; therefore, these values are not expected to have an actual physical meaning.

Changing these outlet parameters for Detroit Dam did somewhat affect modeled water temperatures within Detroit Lake. A tradeoff was made between water temperature errors in Detroit Lake and water temperature errors downstream. Goodness-of-fit statistics for the updated Detroit Lake model are compared to those of the original model in table A1. The mean error (ME) is the sum of the differences between modeled and measured temperatures, where they coincide in space and time, and is an overall measure of bias; a ME close to zero is desirable. The mean absolute error (MAE) is the average of the absolute value of modeled-measured differences and represents a typical error for any data point; an MAE less than 1.0°C has been noted in previous model applications as a reasonable metric denoting a good fit to the data (Sullivan and others, 2007). The root mean square error (RMSE) is the square root of the average squared error between modeled-measured data comparisons and is equal to the square of the mean plus the square of the standard deviation. If the ME is zero, then the RMSE is equal to the standard deviation of the errors—a good measure of the magnitude of the typical error of the prediction; RMSE values less than 1.0 to 1.5°C have been deemed a good fit in previous applications.

Final Big Cliff Modeled Water Temperature

Changing the outlet setup of the Detroit Lake model outlets provided a better match between the Big Cliff Dam release temperatures and the measured water temperatures at Niagara (fig. A3). Agreement with those measured data downstream was good, with a mean absolute error less than 0.4°C in both 2002 and 2003 (table A2). The construction and calibration of the Big Cliff Reservoir model now allows the Detroit Lake model to be connected with the existing North Santiam River model and other Willamette River basin models downstream.

Table A1. Detroit Lake model goodness-of-fit statistics for calendar years 2002 and 2003 for the original Detroit Lake model (Sullivan and others, 2007) and the updated Detroit Lake model used as the upstream boundary for the Big Cliff Reservoir model.

Detroit Lake model	Year 2002		Year 2003	
	Original model	**Updated model**	**Original model**	**Updated model**
Mean error	-0.02	-0.39	-0.34	-0.55
Mean absolute error	0.52	0.77	0.58	0.77
Root mean square error	0.69	1.00	0.76	0.99

Table A2. Big Cliff model goodness-of-fit statistics for calendar years 2002 and 2003 using the updated Detroit Lake model as the upstream boundary condition.

Statistic	Year 2002	Year 2003
Mean error	-0.05	-0.09
Mean absolute error	0.31	0.39
Root mean square error	0.39	0.48

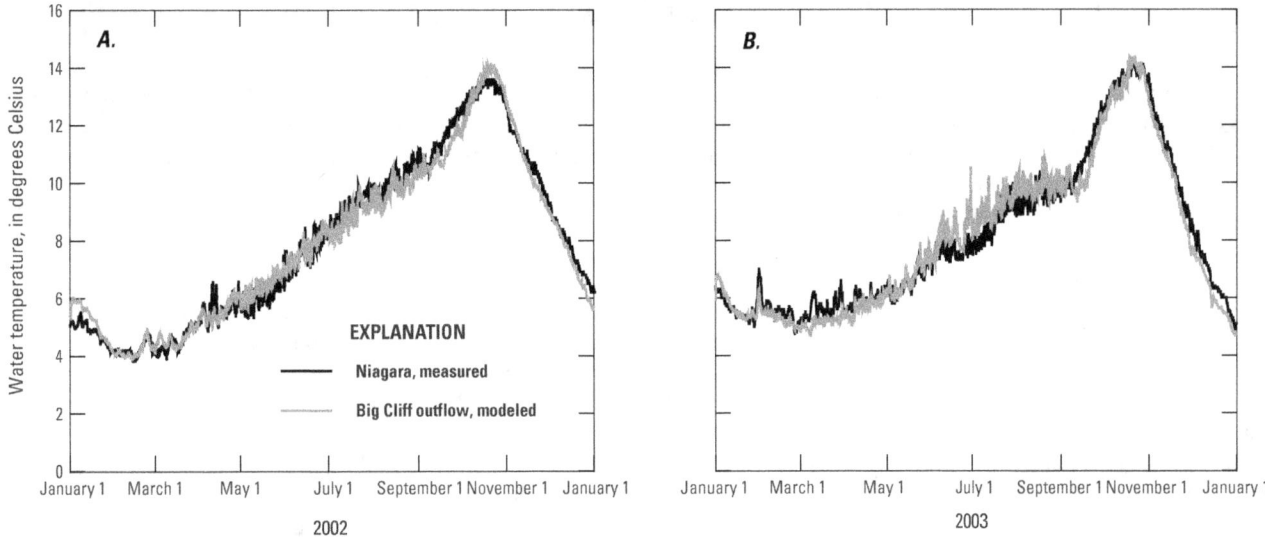

Figure A3.   Modeled water temperatures released from Big Cliff Dam compared to measured water temperatures in the North Santiam River at Niagara, Oregon, 0.7 mile downstream.

Appendix B.   Detroit Lake and Big Cliff Reservoir Model Evaluation for 2011

To ensure that the CE-QUAL-W2 models of Detroit Lake and Big Cliff Reservoir would accurately represent conditions resulting from current dam operations, the models were tested using observed conditions from 2011. Since 2007, operations at Detroit Dam have expanded to include more frequent use of the spillway (elevation 1,541.0 ft; 469.7 m) and the upper regulating outlet (RO, elevation 1,339.9 ft; 408.4 m) to improve downstream temperature management. To better match measured temperatures at the USGS Niagara gaging station (14181500), the CE-QUAL-W2 model line width parameter for the spillway outlet, which was not used in the original model calibration for 2002–03, was set to 25 m through an optimization process. A sensitivity analysis of the line width of the upper RO at Detroit Dam also was conducted, but resulted in little difference in simulated outflow temperatures; therefore, the line width for the ROs was left at 4.0 m. The line width used for the power penstocks was unchanged and remained at 6.8 m.

Comparisons of modeled and measured vertical temperature profiles within Detroit Lake (fig. B1) and Big Cliff Reservoir (fig. B2) show that the models capture the seasonal patterns in the vertical profiles relatively well, with perhaps a slight negative bias for the deepest profiles. The Detroit Lake model also does not capture some of the daily variation in the mid-depth temperature profile data. Modeled daily mean release temperatures from Detroit and Big Cliff Dams (fig. B3) show that significant heat exchange is occurring in Big Cliff Reservoir in August and September; including the Big Cliff Reservoir model, therefore, is useful for capturing these heat-exchange processes. The comparison of Big Cliff Dam modeled release temperatures to measured temperatures downstream at the Niagara gage (fig. B3) shows relatively good agreement with an MAE less than 1.0°C, but a slight negative bias for August through December. A similar comparison between Big Cliff Dam modeled hourly release temperatures and measured hourly temperatures at Niagara is shown in figure B4.

Goodness-of-fit statistics are noted on the figures in this appendix to quantify the overall model performance. Definitions of the mean error (ME), mean absolute error (MAE), and the root mean square error (RMSE) were noted in appendix A. The Nash-Sutcliffe coefficient (NS) is the proportion of variance in the measured values that is explained by the predicted values, and is a more rigorous fit statistic than the coefficient of determination. An NS value of 1.0 represents a perfect fit, an NS value of 0 indicates that the model predictions are only as accurate as the mean of the measured data, and an NS value less than zero means that the measured mean is a better predictor than the model (Nash and Sutcliffe, 1970). In this case, the NS values are all roughly 0.9 or higher, indicating that the model captures most of the variance in the measured data.

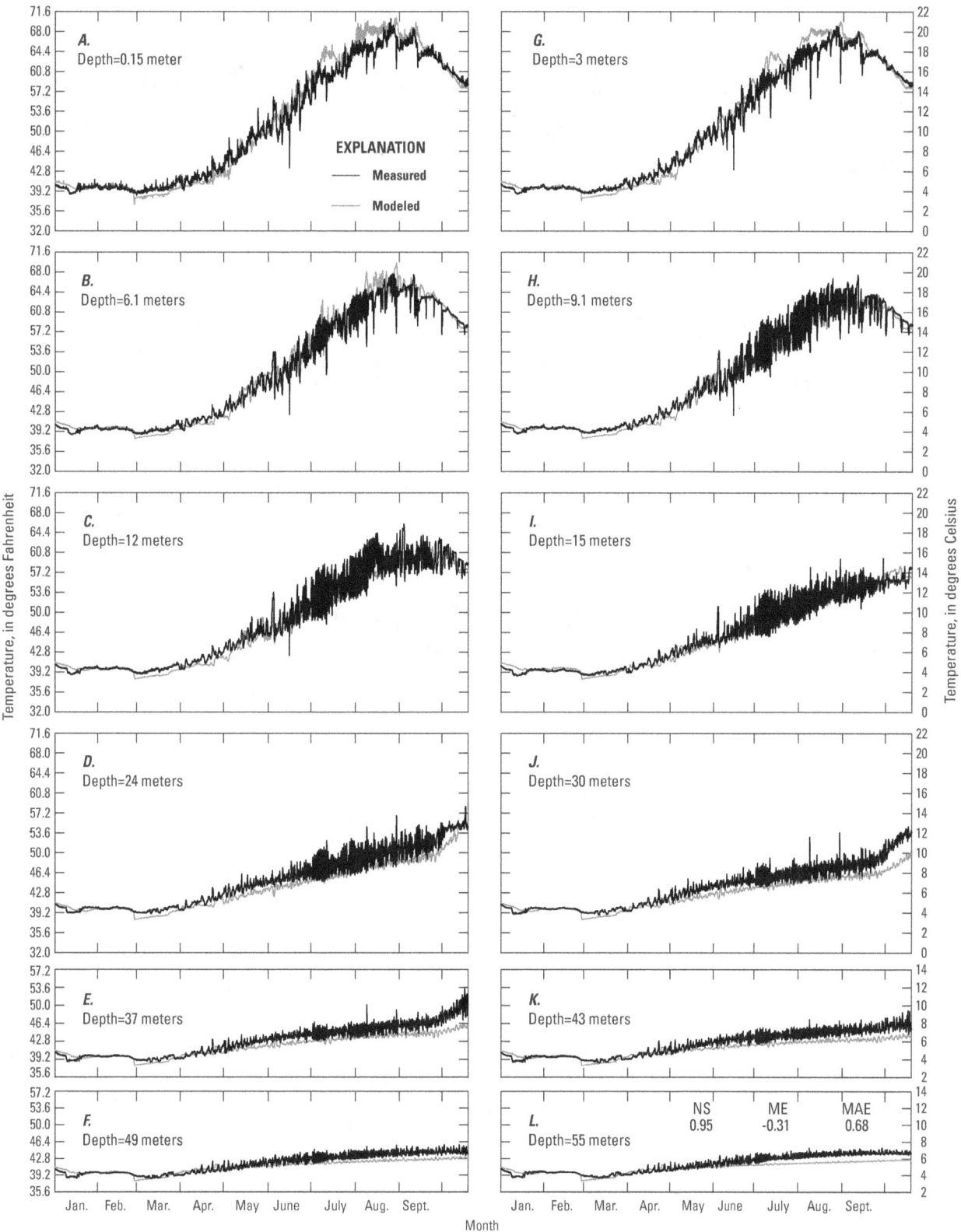

Figure B1.   Measured and modeled water temperatures in Detroit Lake, Oregon, at discrete depths within the lake during 2011. NS is the Nash-Sutcliffe coefficient, ME is the mean error, and MAE is the mean absolute error between the measured and modeled water temperature.

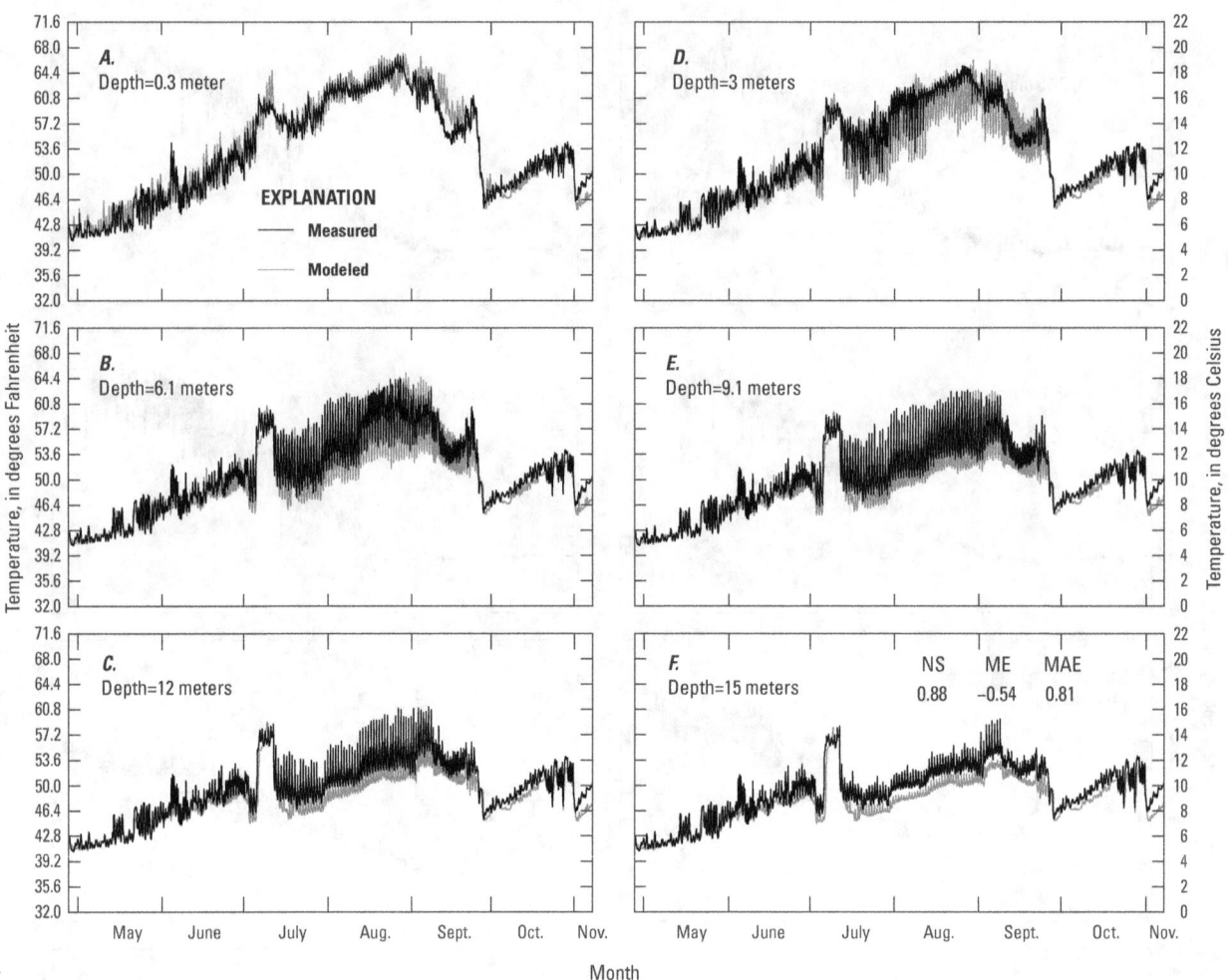

Figure B2.   Measured and modeled water temperatures in Big Cliff Reservoir, Oregon, at discrete depths within the lake during 2011. NS is the Nash-Sutcliffe coefficient, ME is the mean error, and MAE is the mean absolute error between the measured and modeled water temperature.

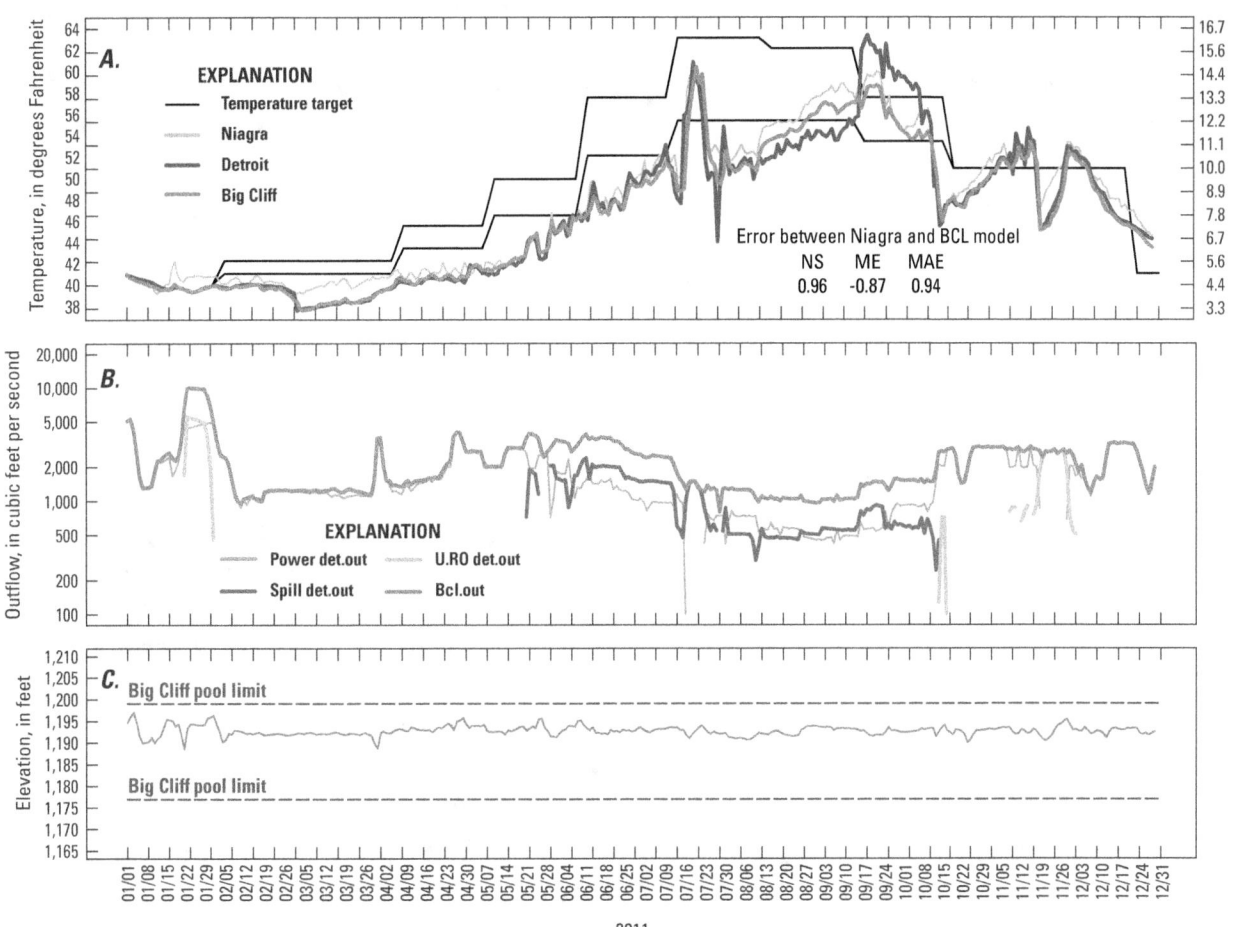

Figure B3. Simulated daily mean water temperatures from Detroit Dam release, Big Cliff Dam release, and measured daily mean temperatures from USGS gaging station 14181500 (North Santiam River at Niagara, Oregon), during 2011. NS is the Nash-Sutcliffe coefficient, ME is the mean error, and MAE is the mean absolute error between the measured and modeled water temperature.

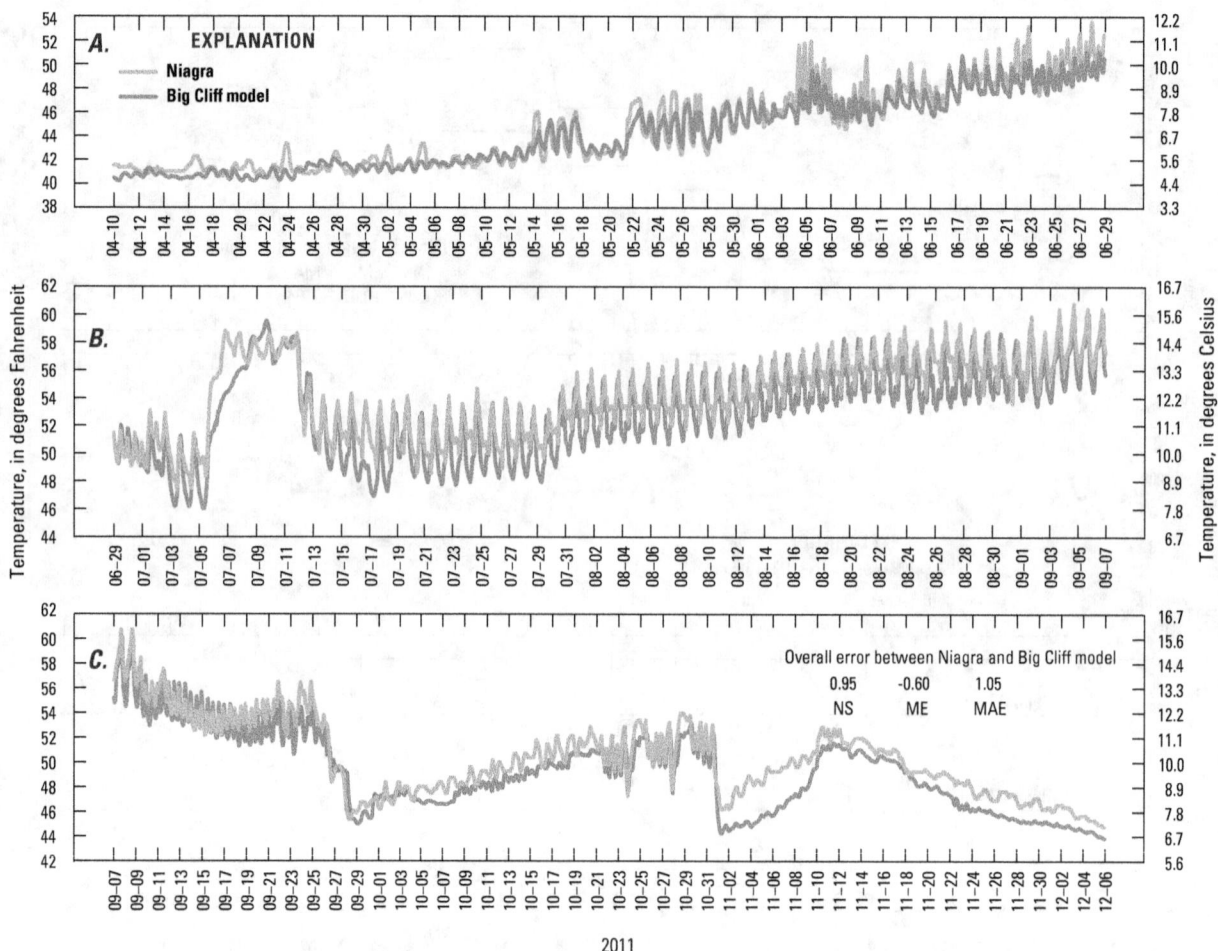

Figure B4.   Simulated hourly water temperatures from modeled Big Cliff Dam, Oregon, releases during 2011, compared to measured hourly temperatures from USGS gaging station 14181500 (North Santiam River at Niagara).

Appendix C.   Code Modifications

Several modifications were made to the version 3.12 CE-QUAL-W2 model code beyond those already documented by Sullivan and others (2007) and Sullivan and Rounds (2006). The original Detroit Lake version 3.12 model published by Sullivan and others (2007) included a USGS-coded blending subroutine that allowed the model to select two outlets from among several potential outlets and set the percentage of the total release rate in each outlet in order to meet a user-supplied downstream temperature target time series. In this study, several additional modifications were made to the model code:

- The DOWNSTREAM_WITHDRAWAL and LATERAL_WITHDRAWAL subroutines were updated to make the velocity profile equations similar to those used in CE-QUAL-W2 version 3.6. The LATERAL_WITHDRAWAL subroutine also was modified to allow both "point" and "line" withdrawals, using equations from the DOWNSTREAM_WITHDRAWAL subroutine.

- The blending subroutine was modified to iteratively set the outlet release rates when two outlets are used; up to five iterations are performed. In the original blending subroutine, the temperature released from each outlet was estimated as the temperature in the reservoir at the elevation of the outlet, but that estimate did not always match well with the actual temperature released through the outlet because water from many layers is drawn into the outlet as specified by the model's selective withdrawal algorithm. Because the amount of water that comes from various layers depends on the release rate for that outlet, this procedure had to be iterative in order to be accurate.

  In this iterative process, an initial estimate of the release rates is made and then refined in a loop. The loop is exited early if the current and previous release rates for each outlet are within 1 percent. If all five iterations are performed and the current and previous results are still not within 1 percent, then a message is written to the model warning file. This code modification greatly improved the accuracy of the blending subroutine.

- Two new inputs were added to the model control file to allow withdrawals (in addition to structures) to have the characteristics of a line sink or a point sink and to specify the width of the line sink. The WD SINK card is located in the control file just after the WD TYPE card; accepted inputs are either LINE or POINT. The WD WIDTH card follows, to include the width of any line sinks that are specified for withdrawals.

- For floating outlets, the original blending algorithm assumed that all floating outlets, and all sliding-gate outlets that were located at their upper limit near the water surface, were positioned at a depth of 1.5 m below the water surface. The introduction of a new withdrawal depth variable allows the user to set the depth of each floating or sliding-gate outlet relative to the water surface. This new WD FLOAT card is located just after the WD WIDTH card in the control file.

- A new input variable (MINFRAC) was added for use with the blending subroutine to allow the user to specify a minimum outlet flow rate or a minimum fraction of the total release rate for a particular group of withdrawals. This input variable does not affect the choice of outlets. If a user specifies more than three withdrawals in any one withdrawal group, this factor is not used to determine which two of the available withdrawals are selected for blending the releases and, therefore, this constraint will not necessarily be honored.

  When not active, this new MINFRAC variable is set to 0.0. When used to set a minimum fraction of the total withdrawal, it is set to a positive value less than or equal to 1.0. To set a minimum flow rate in cubic meters per second, the MINFRAC input is set by the user to a negative number where the minimum flow rate is equal to the absolute value of MINFRAC. The negative sign is used to tell the model that this is a minimum flow rate rather than a minimum fraction of the total flow. The model ensures that MINFRAC is not larger than 1.0. This new user input is provided on a new WD MINFR card that is located just after the WD GROUP card in the model control file.

  Note that the minimum flow specification should work well as long as the withdrawn flows do not change between the times that the blending subroutine is called. If the withdrawal flow rates change more frequently than the subroutine is called, then this code does not guarantee that the minimum specified flows are met. Often, the blending subroutine may only be called once or a few times per day so that dam operators do not have to frequently change gate positions.

- Lastly, a withdrawal priority setting was added as a user-specified input so that if more than one minimum flow constraint was specified within a withdrawal group, the code can try to honor each constraint according to the specified outlet priority. A lower code specifies a higher priority; therefore, a priority input of 1 is a higher priority than a priority input of 2. Specified minimum flow criteria will not necessarily be met if the outlet has a lower priority (higher priority number). The priority setting does not affect the choice of which two outlets are selected from among a group of more than two withdrawals in a group.

After honoring the priority level, the blending subroutine makes an attempt to meet any other fractional flow or minimum flow criteria. However, if a greater total flow or total fractional flow is requested relative to what can be provided, a compromise will be made and it is possible that neither minimum flow constraint will be honored.

In cases when the water column is isothermal and no minimum flow criteria are set, the priority setting is used to route all of the outflow to the outlet with the higher priority (lower priority setting).

The new inputs to the model control file described in this section above might look like the snippet of a control file example as reproduced here for a system with three withdrawals:

WD TYPE	WDTYPE FIXED	WDTYPE FIXED	WDTYPE FIXED	WDTYPE	WDTYPE	WDTYPE	WDTYPE	WDTYPE	WDTYPE
WD SINK	WDSINK LINE	WDSINK LINE	WDSINK LINE	WDSINK	WDSINK	WDSINK	WDSINK	WDSINK	WDSINK
WD WIDTH	WWD 6.8	WWD 4.0	WWD 25	WWD	WWD	WWD	WWD	WWD	WWD
WD FLOAT	WDEPTH 0	WDEPTH 0	WDEPTH 0	WDEPTH	WDEPTH	WDEPTH	WDEPTH	WDEPTH	WDEPTH
WD GROUP	WDGRP 1	WDGRP 1	WDGRP 2	WDGRP	WDGRP	WDGRP	WDGRP	WDGRP	WDGRP
WD MINFR	MINFRAC 0.4	MINFRAC 0	MINFRAC 0	MINFRAC	MINFRAC	MINFRAC	MINFRAC	MINFRAC	MINFRAC
WD PRIOR	WDPRIOR 1	WDPRIOR 2	WDPRIOR 3	WDPRIOR	WDPRIOR	WDPRIOR	WDPRIOR	WDPRIOR	WDPRIOR

The modified model source code is available online at the project website: http://or.water.usgs.gov/santiam/detroit_lake/ in the Downloads section of that page.

Appendix D.   North Santiam and Santiam River Model Set-Up and Application

The original North Santiam and Santiam River model was constructed and calibrated for June through October 2001 and April through October 2002 (Sullivan and Rounds, 2004). For application to the conditions of this study, updated flow, temperature, and meteorological boundary conditions had to be created.

Boundary Inflows

Boundary inflows to the river model include the North Santiam River below Big Cliff Dam (RM 58.1), Rock Creek (RM 49.3), the Little North Santiam River (RM 39.2), and the South Santiam River (RM 11.7). Other waters entering the river model include precipitation on the water surface, groundwater inflows, and effluent from the Stayton and Jefferson wastewater treatment plants located at RMs 27.5 and 9.0, respectively.

For the initial simulations representing current conditions and for testing purposes, measured hourly streamflow from the North Santiam River at Niagara (USGS gaging station 14181500) was used as the upstream boundary because that site is near Big Cliff Dam. For simulations in the study based on hypothetical Detroit Dam scenarios, the simulated outflow from the upstream Detroit Lake and Big Cliff Reservoir models was used as the upstream boundary for the river model.

For most of the *normal* and *cool/wet* environmental scenarios, measured hourly streamflow for the Rock Creek near Mill City site (USGS station 14181750) were available and were used directly. For the *hot/dry* scenario, however, no measured data were available for that site. In the absence of measured data, hourly streamflow for Rock Creek was estimated by multiplying the measured hourly streamflow from the Little North Santiam River near Mehama (USGS site 14182500) by a drainage area ratio of 0.17 (ratio of the Rock Creek drainage area to the Little North Santiam River drainage area).

Measured hourly streamflow data were available for the time periods of all three environmental scenarios for the Little North Santiam River (USGS site 14182500). That station is located 2 mi upstream from its confluence with the North Santiam River.

South Santiam River streamflow data were estimated at the river's confluence with the North Santiam River using hourly streamflow measurements at the South Santiam River at Waterloo (USGS site 14187500). Measured streamflows at Waterloo were multiplied by 1.1 (as a drainage area expansion factor) to account for net inflows and withdrawals between the Waterloo gage and the North Santiam River confluence.

Measured precipitation data were obtained from the National Oceanic and Atmospheric Administration's National Climatic Data Center (NCDC). Daily total precipitation data for the upstream two branches of the model, upstream of the Little North Santiam River confluence (RM 39.2), were taken from a site at Detroit Dam (NCDC station 352292). For the downstream three model branches, daily total precipitation data were measured at Stayton, Oregon (NCDC station 358095).

Groundwater inflow, flow from small ungaged creeks, and errors in gaged streamflow data were accounted for by using a distributed tributary in the model for each of the six branches. For the initial simulation, distributed tributary flow files created for 2001–02 by Sullivan and Rounds (2004) were used. These initial inflows were then adjusted in a second simulation for each environmental scenario to eliminate the difference between simulated and measured streamflows in each branch. This adjustment was justified because it could not be assumed that groundwater inflow and the cumulative inflow from small unmeasured creeks would remain similar from year to year.

Additional inflow to the model included municipal wastewater effluent from the cities of Stayton and Jefferson. Records containing daily effluent discharges for both municipalities were provided in monthly Discharge Monitoring Reports (DMRs) to ODEQ (Robert Dicksa, Oregon Department of Environmental Quality, written commun., 2011).

Surface-Water Withdrawals

Surface-water withdrawals from the river model included municipal water supply (cities of Gates, Mill City, Salem, Stayton, and Jefferson) and irrigation (Santiam Water Control District and the Sidney Irrigation Cooperative) (fig. 3). Data for these withdrawals were provided by the Oregon Water Resources Department Water Use Reporting Database. Although the withdrawals were reported as monthly total volumes, they were converted to a rate in cubic meters per second for the model. Monthly withdrawal rates were assigned to just the midpoint of each month, and the model then was set up to linearly interpolate between those monthly midpoints for each model time step.

Boundary Water Temperatures

Using the same methods employed by Sullivan and Rounds (2004), measured or estimated water temperatures were assigned to all boundary flows entering the river model. For testing purposes and for simulating existing conditions, hourly water temperature data measured in the North Santiam River at Niagara (USGS station 14181500) were used as the upstream boundary for the model at Big Cliff Dam. For other model scenarios, simulated release temperatures from the Big Cliff Reservoir model were used.

Measured hourly water temperature data were available for the Little North Santiam River near Mehama (USGS site 14182500). Measured hourly water temperature data from Rock Creek near Mill City (USGS site 14181750) were available only for 2006. Although Sullivan and Rounds (2004) used measured water temperature data from the Little North Santiam River near Mehama as estimated water temperatures for Rock Creek in their 2001–02 simulations, a comparison of water temperatures in Rock Creek and the Little North Santiam River showed that Little North Santiam River water temperatures were about 10°C warmer than temperatures in Rock Creek in summer. Because of that difference, measured hourly water temperature data from 2006 for the Rock Creek site were used for all three (*hot/dry*, *normal*, and *cool/wet*) environmental scenarios.

Measured daily-mean air temperature data for precipitation entering the river model through the water surface were available from the Detroit and Stayton NCDC meteorological stations that were used for the daily precipitation data. Data from the Detroit station were used for branches 1–2, and data from the Stayton station were used for branches 3–6.

Water temperatures for groundwater and ungaged tributaries flowing into the six model branches were estimated using a weighted average approach from Sullivan and Rounds (2004). For the two upstream model branches, 70 percent of the flow was assumed to come from ungaged tributaries. Daily-mean water temperatures measured at the Little North Santiam River near Mehama (USGS site 14182500) were used for this portion of the flow. For the remaining 30 percent of flow, which was assumed to be groundwater, a constant temperature of 11.5°C was assigned. For the four downstream model branches, flow was weighted as 50 percent ungaged tributary flow and 50 percent groundwater and the temperatures were estimated accordingly.

For the Stayton and Jefferson municipal wastewater outflows, measured daily-mean water temperatures were taken from information included in the monthly DMRs provided to ODEQ.

Meteorological Data

Meteorological inputs for the three environmental scenarios for the river model were constructed using the same data sources and methods as those used by Sullivan and Rounds (2004). Air temperature, dew-point temperature, wind speed, and wind direction data were obtained from the Stayton meteorological site. Solar radiation data were obtained from the Solar Radiation Monitoring Laboratory at the University of Oregon for their meteorological site in Eugene. Cloud cover data were computed using a comparison between measured and theoretical clear-sky solar radiation rates, in which the nighttime cloud cover information was interpolated from values near sunrise and sunset. For more information on these data sources and methods, see Sullivan and Rounds (2004).

Appendix E.   Additional Model Scenarios

Some of the model scenarios included in this study resulted in temperature releases from Detroit Dam that were either not very successful in matching the intended downstream target or closely matched those of another scenario. This appendix is dedicated to the archival of that set of model scenarios.

Late Refill of Detroit Lake

Delaying the time at which Detroit Lake begins to fill from February 1 to June 1 results in the operational scenario *late_refill* (fig. E1, tables 4 and 5). The thinking behind

this scenario was that by keeping the lake level low during spring, it might be easier or more efficient to pass juvenile fish downstream past the dam. Insufficient streamflow was available to fill the lake after June 1, and the lake level remained well below full pool level for the rest of the summer under all three environmental scenarios. Temperatures released from existing outlets under this late-refill scenario generally did not meet the *max* temperature targets for much of the year (fig. E2). Early in the summer, power production constraints and a water level that was too low to use the spillway resulted in water releases that were cooler than the target. In autumn, the lake level was too low for existing outlets to access sufficient cool water, and the water releases were too warm.

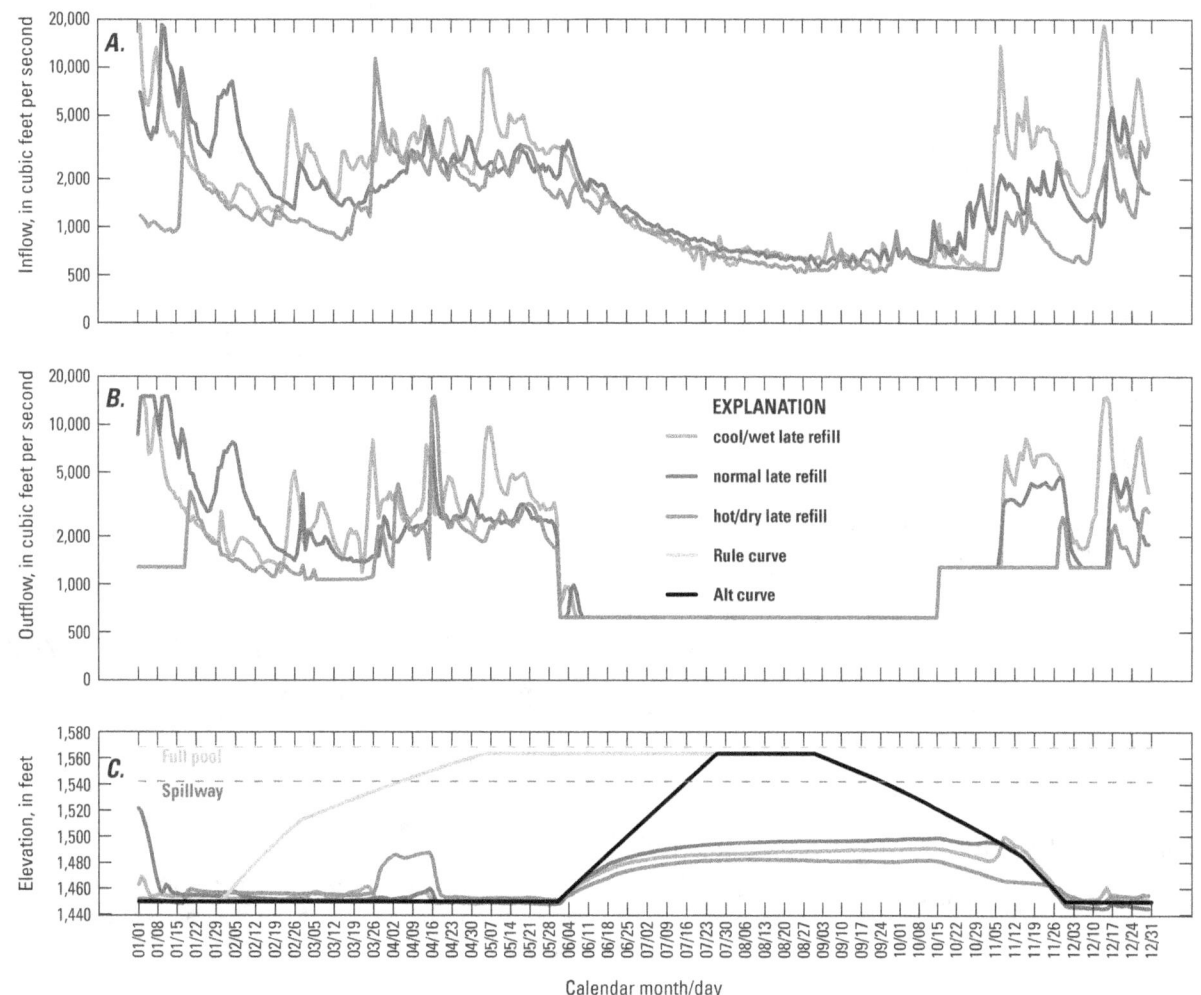

Figure E1.   Comparison of *late_refill* operational scenarios (*c4*, *n4*, *h4*): (*A*) total inflows, (*B*) total outflows, and (*C*) modeled water-surface elevation.

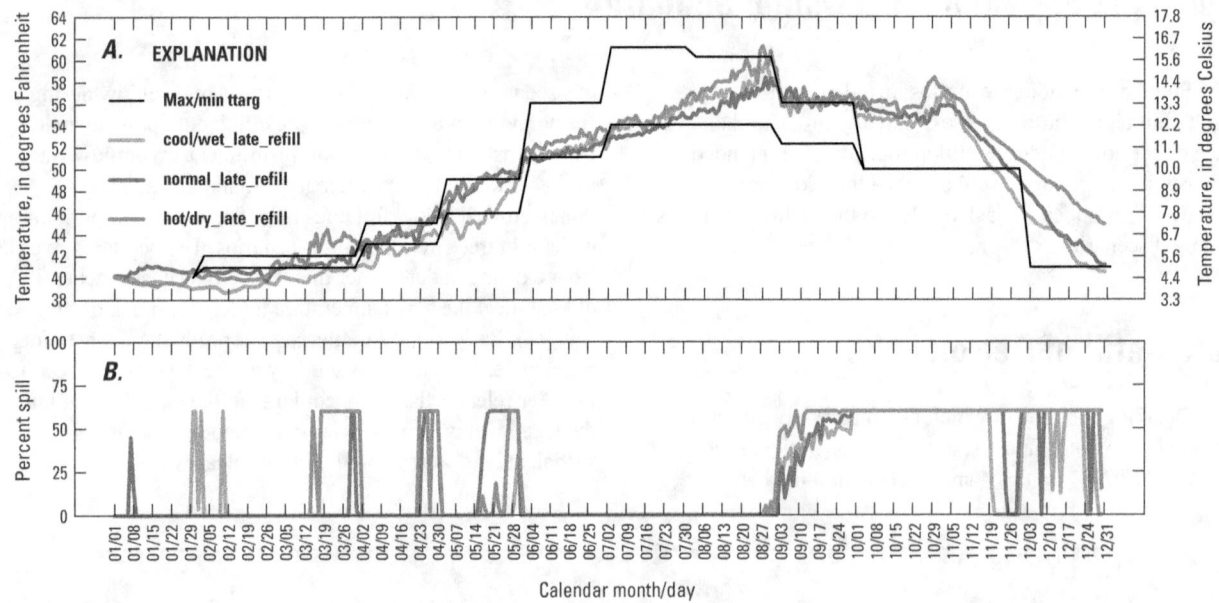

Figure E2.   (*A*) Modeled water temperature and (*B*) percent spill for *existing* structural scenarios (*c4, n4, h4*) with *late_refill* operations and *max* temperature targets. The maximum and minimum temperature target established for the McKenzie River (labeled "Max/min ttarg") is shown but only the maximum was used in this simulation.

Early Drawdown of Detroit Lake

Changing the time at which the Detroit Lake level reaches its minimum conservation pool from December 1 to November 1 results in the operational scenario *early_dd* (tables 4, 5, and fig. E3). Despite the earlier drawdown, this operational scenario had little effect on the date at which the use of the spillway was no longer an option, as the lake level under base operations tended to decrease to the elevation of the spillway crest before this change occurred. However, the earlier drawdown caused the thermocline to be drawn down to the level of the available outlets sooner than in base operations, which caused the *max* temperature target generally not to be met during October and November (fig. E4).

Power Penstocks with Floating Outlet and 10 Percent Minimum Power Generation

Several model scenarios were run with the *pp-float* structural option and a combination of flow constraints. In that group, this scenario specified that a minimum of 10 percent

of the total outflow be released through the power penstocks (table 7), providing another point in a continuum between a scenario with no power generation constraint (*c10, n10, h10*) and a scenario with a 20 percent power generation minimum (*c12, n12, h12*). Used with the *max* temperature targets, the results did not meet the target release temperatures in October and November for the *hot/dry* environmental scenario (compare fig. E5 with figs. 21 and 22).

Sliding-Gate and Floating Outlets with Delayed Drawdown

The combination of a sliding-gate outlet and a floating outlet (*slider1340-float*) along with delayed drawdown operations (*delay_dd2*) provided another combination that was of interest and similar to a couple of scenarios documented in the main part of this report (table 7). Results from *c18, n18,* and *h18* are almost identical to results from *uro-float_400fmin* scenarios *c14, n14,* and *h14* (fig. E6).

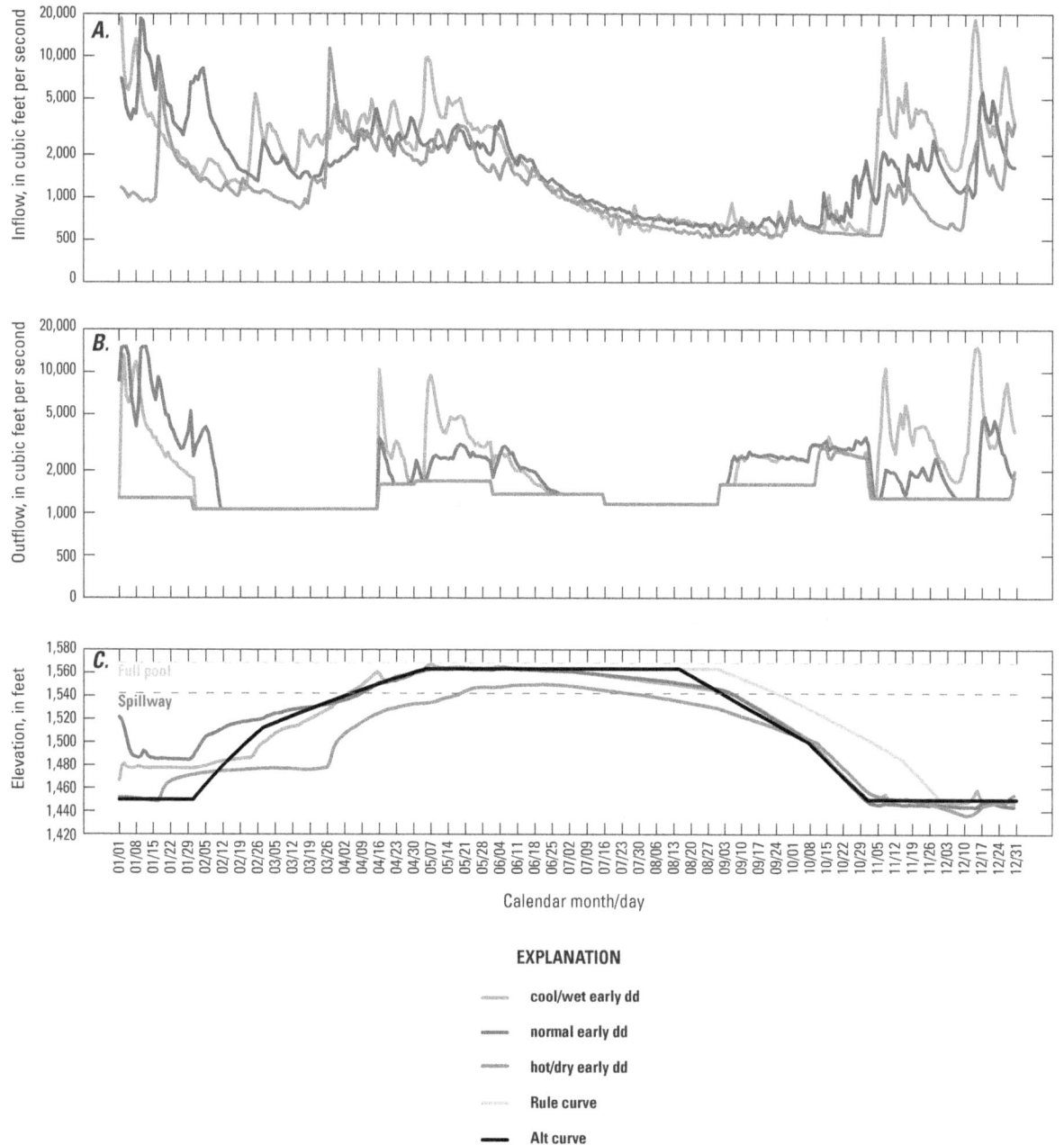

Figure E3. Comparison of *early_dd* operational scenarios (*c5, n5, h5*): (*A*) total inflows, (*B*) total outflows, and (*C*) modeled water-surface elevation.

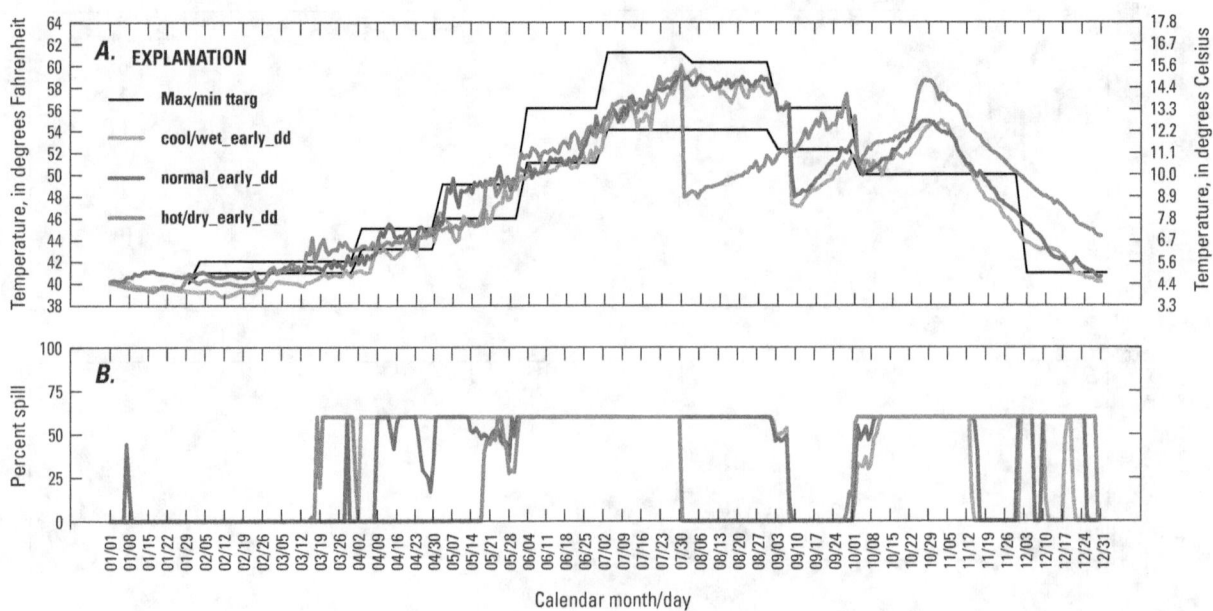

Figure E4.   (*A*) Modeled water temperature and (*B*) percent spill for *existing* structural scenarios (*c5*, *n5*, *h5*) with *early_dd* operations and *max* temperature targets. The maximum and minimum temperature target established for the McKenzie River (labeled "Max/min ttarg") is shown but only the maximum was used in this simulation.

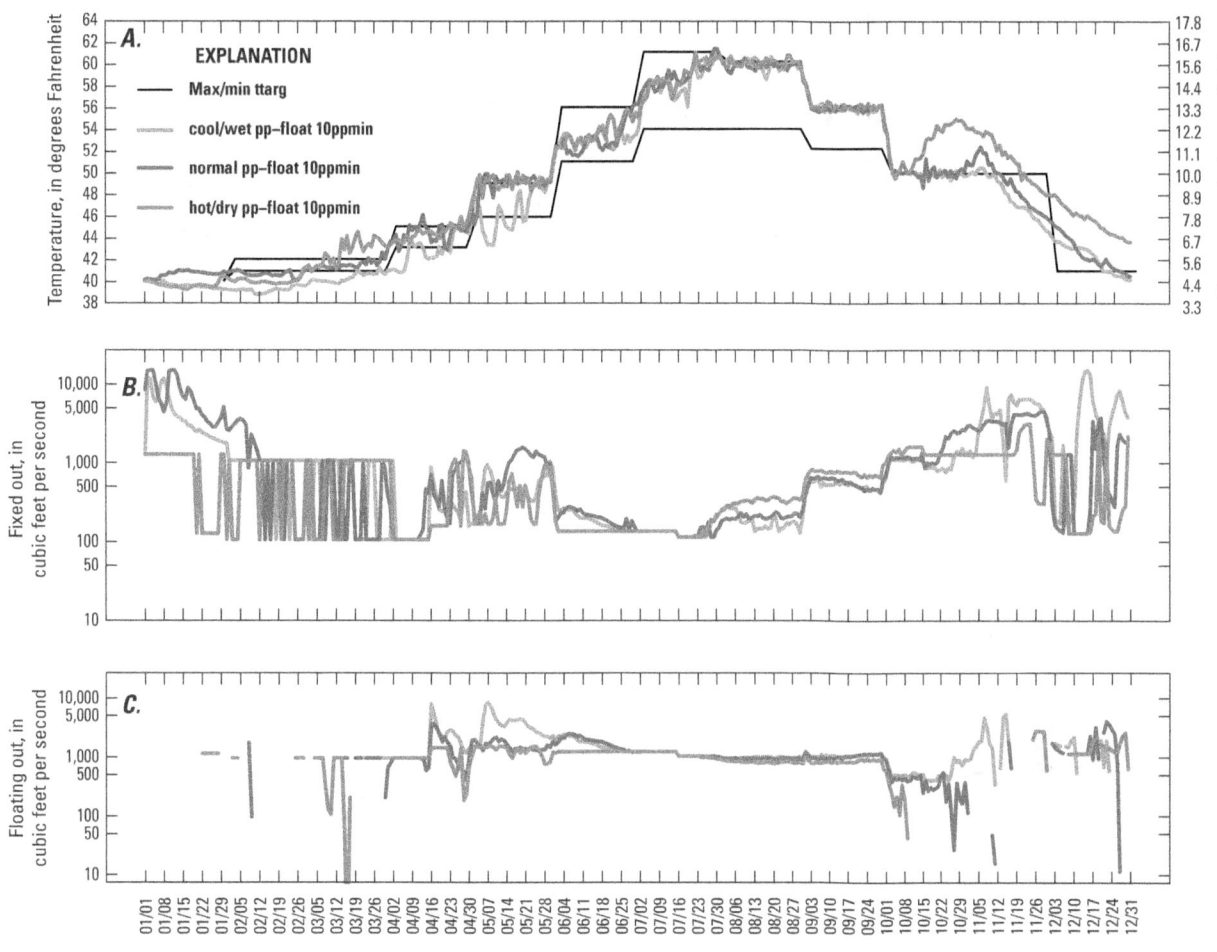

Figure E5.   Results from *pp-float* structural scenarios with *10ppmin* operations and *max* temperature targets (scenarios *c11, n11, h11*): (*A*) modeled water temperature, (*B*) outflow from fixed outlet, and (*C*) outflow from floating outlet. The maximum and minimum temperature target established for the McKenzie River (labeled "Max/min ttarg") is shown but only the maximum was used in this simulation.

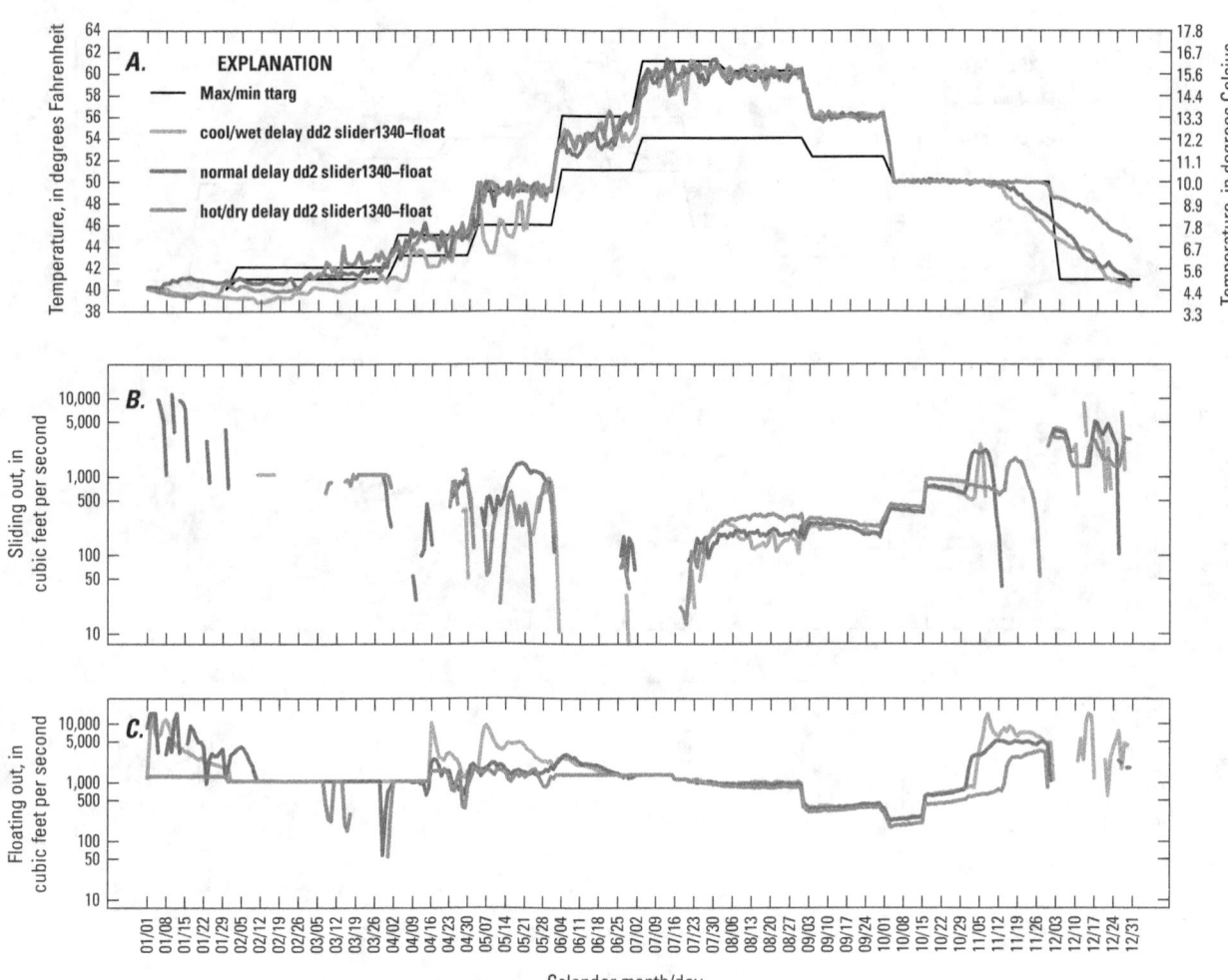

Figure E6.   Results from *slider1340-float* structural scenarios with *delay_dd2* operations and *max* temperature targets (scenarios *c18*, *n18*, *h18*): (*A*) modeled water temperature, (*B*) outflow from sliding outlet, and (*C*) outflow from floating outlet. The maximum and minimum temperature target established for the McKenzie River (labeled "Max/min ttarg") is shown but only the maximum was used in this simulation.

www.ingramcontent.com/pod-product-compliance
Lightning Source LLC
Chambersburg PA
CBHW081601170526
45166CB00009B/2778